Soil Fertility, Fertilizer and Integrated Nutrient Management

Soil Fertility, Fertilizer and Integrated Nutrient Management

Ravinder Kumar Kashyap

Oxford Book Company
Jaipur ▪ New Delhi

Published By

Oxford Book Company

Regd. Off. "Bony Residency", Gate No. 2
Opp. Tilak Public School, Vishweswariya Nagar,
Gopalpura Road, Jaipur - 302018, India
Ph: 0141 - 2761280, Telefax: 2761381
email - oxfordbookjpr@gmail.com
website - oxfordbookcompany.com

Delhi Office: 102, 1st Floor, Satyam House,
4327/3, Ansari Road, Darya Ganj,
New Delhi - 110002, India
Ph: 011-45652440

Distributed by
KEYA INTERNATIONAL PRESS LLP
E-mail: keyainternationalpressllp@gmail.com

Published in India, 2026

ISBN : 9788189473594

Typeset by Shivangi Computers, Jaipur

PRINTED IN INDIA BY

Global Printing Services,
116, 1st Floor F.I.E. Patparganj Industrial Area, New Delhi-110092.

Contents

Preface

Soil, also commonly referred to as earth or dirt, is a mixture of organic matter, minerals, gases, liquids, and organisms that together support the life of plants and soil organisms. Some scientific definitions distinguish dirt from soil by restricting the former term specifically to displaced soil. A typical soil is about 50% solids (45% mineral and 5% organic matter), and 50% voids (or pores) of which half is occupied by water and half by gas. Bioavailable phosphorus (available to soil life) is the element in soil that is most often lacking. Nitrogen and potassium are also needed in substantial amounts. For this reason these three elements are always identified on a commercial fertilizer analysis. For example, a 10-10-15 fertilizer has 10 percent nitrogen, 10 percent available phosphorus (P_2O_5) and 15 percent water-soluble potassium (K_2O). Sulfur is the fourth element that may be identified in a commercial analysis—e.g. 21-0-0-24 which would contain 21% nitrogen and 24% sulfate.

Soil fertility is a complex process that involves the constant cycling of nutrients between organic and inorganic forms. As plant material and animal wastes are decomposed by micro-organisms, they release inorganic nutrients to the soil solution, a process referred to as mineralization. Those nutrients may then undergo further transformations which may be aided or enabled by soil micro-organisms. Like plants, many micro-organisms require or preferentially use inorganic forms of nitrogen, phosphorus or potassium and will compete with plants for these nutrients, tying up the nutrients in microbial biomass, a process often called immobilization. The balance between immobilization and mineralization processes depends on the balance and availability of major nutrients and organic carbon to soil microorganisms. Natural processes such as lightning strikes may fix atmospheric nitrogen by converting it to (NO_2). Denitrification may occur under anaerobic conditions (flooding) in the presence of denitrifying bacteria. Nutrient cations, including potassium and many micronutrients, are held in relatively strong bonds with the negatively charged portions of the soil in a process known as cation exchange. Inorganic fertilizers are generally less expensive and have higher concentrations of nutrients than organic fertilizers. Also, since nitrogen, phosphorus and potassium generally must be in the inorganic forms to be taken up by plants, inorganic fertilizers are generally immediately bioavailable to plants without modification. However, some have criticized the use of inorganic fertilizers, claiming that the water-soluble nitrogen doesn't provide for the long-term needs of the plant and creates water pollution. Slow-release fertilizers may reduce leaching loss of nutrients and may make the nutrients that they provide available over a longer period of time.

This book collects and discusses information on production and management of organic manures, biofertilizers, integrated nutrient management in cropping systems. The book will be useful reading for students, teachers and all related to agriculture fields.

— *Ravinder Kumar Kashyap*

Chapter 1

Introduction to Soil

What is Soil

Soil, also commonly referred to as earth or dirt, is a mixture of organic matter, minerals, gases, liquids, and organisms that together support the life of plants and soil organisms. Some scientific definitions distinguish dirt from soil by restricting the former term specifically to displaced soil.

Figure: *Surface-water-gley developed in glacial till in Northern Ireland*

Soil consists of a solid phase of minerals and organic matter (the soil matrix), as well as a porous phase that holds gases (the soil atmosphere) and water (the soil solution).

Accordingly, soil is a three-state system of solids, liquids, and gases. Soil is a product of several factors: the influence of climate, relief (elevation, orientation, and slope of terrain), organisms, and the soil's parent materials (original minerals) interacting over time. It continually undergoes development by way of numerous physical, chemical and biological processes, which include weathering with associated erosion. Given its complexity and strong internal connectedness, soil ecologists regard soil as an ecosystem. Most soils have a dry bulk density (density of soil taking into account voids when dry) between 1.1 and 1.6 g/cm, though the soil particle density is much higher, in the range of 2.6 to 2.7 g/cm^3.

Figure: *Soil measuring and surveying device*

Little of the soil of planet Earth is older than the Pleistocene and none is older than the Cenozoic, although fossilized soils are preserved from as far back as the Archean. Collectively the Earth's body of soil is called the pedosphere. The pedosphere interfaces with the lithosphere, the hydrosphere, the atmosphere, and the biosphere. Soil has four important functions:

- as a medium for plant growth
- as a means of water storage, supply, and purification
- as a modifier of Earth's atmosphere
- as a habitat for organisms

All of these functions, in their turn, modify the soil and its properties.

Soil science has two basic branches of study: edaphology and pedology. Edaphology studies the influence of soils on living things. Pedology focuses on the formation, description (morphology), and classification of soils in their natural environment. In engineering terms, soil is included in the broader concept of regolith, which also includes other loose material that lies above the bedrock, as can be found on the Moon and other celestial objects.

Processes

Soil is a major component of the Earth's ecosystem. The world's ecosystems are impacted in far-reaching ways by the processes carried out in the soil, with effects ranging from ozone depletion and global warming to rainforest destruction and water pollution. With respect to Earth's carbon cycle, soil acts as an important carbon reservoir, and it is potentially one of the most reactive to human disturbance and climate change. As the planet warms, it has been predicted that soils will add carbon dioxide to the atmosphere due to increased biological activity at higher temperatures, a positive feedback (amplification). This prediction has, however, been questioned on consideration of more recent knowledge on soil carbon turnover.

Soil acts as an engineering medium, a habitat for soil organisms, a recycling system for nutrients and organic wastes, a regulator of water quality, a modifier of atmospheric composition, and a medium for plant growth, making it a critically important provider of ecosystem services. Since soil has a tremendous range of available niches and habitats, it contains a prominent part of the Earth's genetic diversity.

A gram of soil can contain billions of organisms, belonging to thousands of species, mostly microbial and largely still unexplored. Soil has a mean prokaryotic density of roughly 10 organisms per gram, whereas the ocean has no more than 10 prokaryotic organisms per milliliter (gram) of seawater.

Organic carbon held in soil is eventually returned to the atmosphere through the process of respiration carried out by heterotrophic organisms, but a substantial part is retained in the soil in the form of soil organic matter; tillage usually increases the rate of soil respiration, leading to the depletion of soil organic matter. Since plant roots need oxygen, aeration is an important characteristic of soil.

This ventilation can be accomplished via networks of interconnected soil pores, which also absorb and hold rainwater making it readily available for uptake by plants. Since plants require a nearly continuous supply of water, but most regions receive sporadic rainfall, the water-holding capacity of soils is vital for plant survival.

Soils can effectively remove impurities, kill disease agents, and degrade contaminants, this latter property being called natural attenuation. Typically, soils maintain a net absorption of oxygen and methane and undergo a net release of carbon dioxide and nitrous oxide. Soils offer plants physical support, air, water, temperature moderation, nutrients, and protection from toxins. Soils provide readily available nutrients to plants and animals by converting dead organic matter into various nutrient forms.

Composition

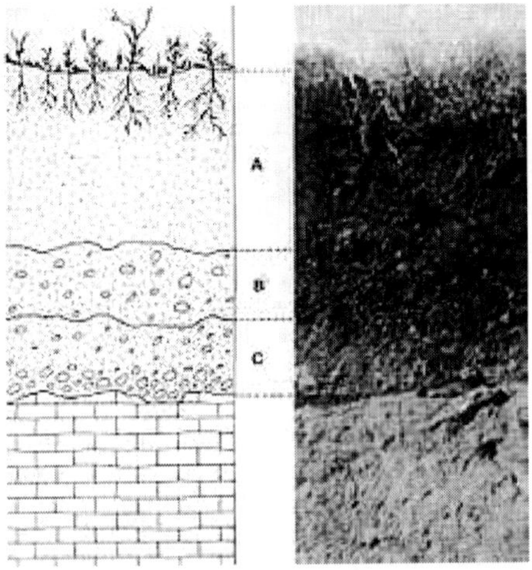

Figure: *A, B, and C represent the soil profile, a notation firstly coined by Vasily Dokuchaev (1846-1903), the father of pedology. Here, A is the topsoil; B is a regolith; C is a saprolite (a less-weathered regolith); the bottom-most layer represents the bedrock.*

A typical soil is about 50% solids (45% mineral and 5% organic matter), and 50% voids (or pores) of which half is occupied by water and half by gas. The percent soil mineral and organic content can be treated as a constant (in the short term), while the percent soil water and gas content is considered highly variable whereby a rise in one is simultaneously balanced by a reduction in the other. The pore space allows for the infiltration and movement of air and water, both of which are critical for life existing in soil. Compaction, a common problem with soils, reduces this space, preventing air and water from reaching plant roots and soil organisms.

Given sufficient time, an undifferentiated soil will evolve a soil profile that consists of two or more layers, referred to as soil horizons. These differ in one or more properties such as in their texture, structure, density, porosity, consistency, temperature, colour, and reactivity. The horizons differ greatly in thickness and generally lack sharp boundaries; their development is dependent on the type of parent material, the processes that modify those parent materials, and the soil-forming factors that influence those processes.

The biological influences on soil properties are strongest near the surface, though the geochemical influences on soil properties increase with depth. Mature soil profiles typically include three basic master horizons: A, B, and C. The solum normally includes the A and B horizons. The living component of the soil is largely confined to the solum, and is generally more prominent in the A horizon. It has been suggested that the pedon, a column of soil extending vertically from the surface to the underlying parent material and large enough to show the characteristics of all its horizons, could be subdivided in the humipedon (the living part, where most soil organisms are dwelling, corresponding to the humus form), the copedon (in intermediary position, where most weathering of minerals takes place) and the lithopedon (in contact with the subsoil).

The soil texture is determined by the relative proportions of the individual particles of sand, silt, and clay that make up the soil.

The interaction of the individual mineral particles with organic matter, water, gases via biotic and abiotic processes causes those particles to flocculate (stick together) to form aggregates or peds. Where these aggregates can be identified, a soil can be said to be developed, and can be described further in terms of colour, porosity, consistency, reaction (acidity), etc.

Water is a critical agent in soil development due to its involvement in the dissolution, precipitation, erosion, transport, and deposition of the materials of which a soil is composed. The mixture of water and dissolved or suspended materials that occupy the soil pore space is called the soil solution. Since soil water is never pure water, but contains hundreds of dissolved organic and mineral substances, it may be more accurately called the soil solution. Water is central to the dissolution, precipitation and leaching of minerals from the soil profile. Finally, water affects the type of vegetation that grows in a soil, which in turn affects the development of the soil, a complex feedback which is exemplified in the dynamics of banded vegetation patterns in semi-arid regions.

Soils supply plants with nutrients, most of which are held in place by particles of clay and organic matter (colloids) The nutrients may be adsorbed on clay mineral surfaces, bound within clay minerals (absorbed), or bound within organic compounds as part of the living organisms or dead soil organic matter. These bound nutrients interact with soil water to buffer the soil solution composition (attenuate changes in the soil solution) as soils wet up or dry out, as plants take up nutrients, as salts are leached, or as acids or alkalis are added.

Plant nutrient availability is affected by soil pH, which is a measure of the hydrogen ion activity in the soil solution. Soil pH is a function of many soil forming factors, and is generally lower (more acid) where weathering is more advanced. Most plant nutrients, with the exception of nitrogen, originate from the minerals that make up the soil parent material. Some nitrogen originates from rain as dilute nitric acid and ammonia, but most of the nitrogen is available in soils as a result of nitrogen fixation by bacteria.

Once in the soil-plant system, most nutrients are recycled through living organisms, plant and microbial residues (soil organic matter), mineral-bound forms, and the soil solution. Both living soil organisms (microbes, animals and plant roots) and soil organic matter are of critical importance to this recycling, and thereby to soil formation and soil fertility. Microbial soil enzymes may release nutrients from minerals or organic matter for use by plants and other microorganisms, sequester (incorporate) them into living cells, or cause their loss from the soil by volatilisation (loss to the atmosphere as gases) or leaching.

Formation

Soil is said to be formed when organic matter has accumulated and colloids are washed downward, leaving deposits of clay, humus, iron oxide, carbonate, and gypsum, producing a distinct layer called the B horizon. This is a somewhat arbitrary definition as mixtures of sand, silt, clay and humus will support biological and agricultural activity before that time. These constituents are moved from one level to another by water and animal activity. As a result, layers (horizons) form in the soil profile. The alteration and movement of materials within a soil causes the formation of distinctive soil horizons. However, more recent definitions of soil embrace soils without any organic matter, such as those regoliths that formed on Mars and analogous conditions in planet Earth deserts.

An example of the development of a soil would begin with the weathering of lava flow bedrock, which would produce the purely mineral-based parent material from which the soil texture forms. Soil development would proceed most rapidly from bare rock of recent flows in a warm climate, under heavy and frequent rainfall. Under such conditions, plants (in a first stage nitrogen-fixing lichens and cyanobacteria then epilithic higher plants) become established very quickly on basaltic lava, even though there is very little organic material. Basaltic minerals commonly weather relatively quickly, according to the Goldich dissolution series. The plants are supported by the porous rock as it is filled with nutrient-bearing water that carries minerals dissolved from the rocks. Crevasses and pockets, local topography of the rocks, would hold fine materials and harbour plant roots. The developing plant roots are associated with mineral-weathering mycorrhizal fungi that assist in breaking up the porous lava, and by these means organic matter and a finer mineral soil accumulate with time. Such initial stages of soil development have been described on volcanoes, inselbergs, and glacial moraines.

How soil formation proceeds is influenced by at least five classic factors that are intertwined in the evolution of a soil: parent material, climate, topography (relief), organisms, and time. When reordered to climate, relief, organisms, parent material, and time, they form the acronym CROPT.

Physical Properties

The physical properties of soils, in order of decreasing importance for ecosystem services such as crop production, are texture, structure, bulk density, porosity, consistency, temperature, colour and resistivity. Soil texture is determined by the relative proportion of the three kinds of soil mineral particles, called soil separates: sand, silt, and clay. At the next larger scale, soil structures called peds or more commonly soil aggregates are created from the soil separates when iron oxides, carbonates, clay, silica and humus, coat particles and cause them to adhere into larger, relatively stable secondary structures.

Soil bulk density, when determined at standardized moisture conditions, is an estimate of soil compaction. Soil porosity consists of the void part of the soil volume and is occupied by gases or water. Soil consistency is the ability of soil materials to stick together. Soil temperature and colour are self-defining. Resistivity refers to the resistance to conduction of electric currents and affects the rate of corrosion of metal and concrete structures which are buried in soil. These properties vary through the depth of a soil profile, i.e. through soil horizons. Most of these properties determine the aeration of the soil and the ability of water to infiltrate and to be held within the soil.

Soil Moisture

Soil water content can be measured as volume or weight. Soil moisture levels, in order of decreasing water content, are saturation, field capacity, wilting point, air dry, and oven dry. Field capacity describes a drained wet soil at the point water content reaches equilibrium with gravity. Irrigating soil above field capacity risks percolation losses. Wilting point describes the dry limit for growing plants. During growing season, soil moisture is unaffected by functional groups or specie richness. Available water capacity is the amount of water held in a soil profile available to plants. As water content drops, plants have to work against increasing forces of adhesion and sorptivity to withdraw water. Irrigation scheduling avoids moisture stress by replenishing depleted water before stress is induced.

Capillary action is responsible for moving groundwater from wet regions of the soil to dry areas. Subirrigation designs (e.g., wicking beds, sub-irrigated planters) rely on capillarity to supply water to plant roots. Capillary action can result in an evaporative concentration of salts, causing land degradation through salination. Soil moisture measurement - measuring the water content of the soil, as can be expressed in terms of volume or weight - can be based on in situ probes (e.g., capacitance probes, neutron probes), or remote sensing methods. Soil moisture measurement is an important factor in determining changes in soil activity.

Soil Gas

The atmosphere of soil, or soil gas, is very different from the atmosphere above. The consumption of oxygen by microbes and plant roots, and their release of carbon dioxide, decreases oxygen and increases carbon dioxide concentration. Atmospheric CO2 concentration is 0.04%, but in the soil pore space it may range from 10 to 100 times that level, thus potentially contributing to the inhibition of root respiration. Calcareous soils regulate CO2 concentration by carbonate buffering, contrary to acid soils in which all CO2 respired accumulates in the soil pore system. At extreme levels, CO2 is toxic. This suggests a possible negative feedback control of soil CO2 concentration through its inhibitory effects on root and microbial respiration (also called soil respiration).

In addition, the soil voids are saturated with water vapour, at least until the point of maximal hygroscopicity, beyond which a vapour-pressure deficit occurs in the soil pore space. Adequate porosity is necessary, not just to allow the penetration of water, but also to allow gases to diffuse in and out. Movement of gases is by diffusion from high concentrations to lower, the diffusion coefficient decreasing with soil compaction. Oxygen from above atmosphere diffuses in the soil where it is consumed and levels of carbon dioxide in excess of above atmosphere diffuse out with other gases (including greenhouse gases) as well as water. Soil texture and structure strongly affect soil porosity and gas diffusion. It is the total pore space (porosity) of soil, not the pore size, and the degree of pore interconnection (or conversely pore sealing), together with water content, air turbulence and temperature, that determine the rate of diffusion of gases into and out of soil. Platy soil structure and soil compaction (low porosity) impede gas flow, and a deficiency of oxygen may encourage anaerobic bacteria to reduce (strip oxygen) from nitrate NO3 to the gases N2, N2O, and NO, which are then lost to the atmosphere, thereby depleting the soil of nitrogen, a detrimental process called denitrification. Aerated soil is also a net sink of methane (CH4) but a net producer of methane (a strong heat-absorbing greenhouse gas) when soils are depleted of oxygen and subject to elevated temperatures.

Soil atmosphere is also the seat of emissions of volatiles other than carbon and nitrogen oxides from various soil organisms, e.g. roots, bacteria, fungi, animals. These volatiles are used as chemical cues, making soil atmosphere the seat of interaction networks playing a decisive role in the stability, dynamics and evolution of soil ecosystems. Biogenic soil volatile organic compounds are exchanged with the aboveground atmosphere, in which they are just 1-2 orders of magnitude lower than those from aboveground vegetation. Humans can get some idea of the soil atmosphere through the well-known 'after-the-rain' scent, when infiltering rainwater flushes out the whole soil atmosphere after a drought period, or when soil is excavated, a bulk property attributed in a reductionist manner to particular biochemical compounds such as petrichor or geosmin.

Solid Phase (Soil Matrix)

Soil particles can be classified by their chemical composition (mineralogy) as well as their size. The particle size distribution of a soil, its texture, determines many of the properties of that soil, in particular hydraulic conductivity and water potential, but the mineralogy of those particles can strongly modify those properties. The mineralogy of the finest soil particles, clay, is especially important.The soil matrix is the solid phase of soils, and comprise the solid particles that make up soils.

Soil particles can be classified by their chemical composition (mineralogy) as well as their size. The particle size distribution of a soil, its texture, determines many of the properties of that soil, in particular hydraulic conductivity and water potential, but the mineralogy of those particles can strongly modify those properties. The mineralogy of the finest soil particles, clay, is especially important.

Gravel, SWand and Silt

Gravel, sand and silt are the larger soil particles, and their mineralogy is often inherited from the parent material of the soil, but may include products of weathering (such as concretions of calcium carbonate or iron oxide), or residues of plant and animal life (such as silica phytoliths). Quartz is the most common mineral in the sand or silt fraction as it is resistant to chemical weathering, except under hot climate; other common minerals are feldspars, micas and ferromagnesian minerals such as pyroxenes, amphiboles and olivines, which are dissolved or transformed in clay under the combined influence of physico-chemical and biological processes.

Mineral Colloids; Soil Clays

Due to its high specific surface area and its unbalanced negative electric charges, clay is the most active mineral component of soil. It is a colloidal and most often a crystalline material. In soils, clay is a soil textural class and is defined in a physical sense as any mineral particle less than 2 -m (8×10 in) in effective diameter. Many soil minerals, such as gypsum, carbonates, or quartz, are small enough to be classified as clay based on their physical size, but chemically they do not afford the same utility as do mineralogically-defined clay minerals. Chemically, clay minerals are a range of phyllosilicate minerals with certain reactive properties.

Before the advent of X-ray diffraction clay was thought to be very small particles of quartz, feldspar, mica, hornblende or augite, but it is now known to be (with the exception of mica-based clays) a precipitate with a mineralogical composition that is dependent on but different from its parent materials and is classed as a secondary mineral. The type of clay that is formed is a function of the parent material and the composition of the minerals in solution. Clay minerals continue to be formed as long as the soil exists. Mica-based clays result from a modification of the primary mica mineral in such a way that it behaves and is classed as a clay. Most clays are crystalline, but some clays or some parts of clay minerals are amorphous. The clays of a soil are a mixture of the various types of clay, but one type predominates.

Typically there are four main groups of clay minerals: kaolinite, montmorillonite-smectite, illite, and chlorite. Most clays are crystalline and most are made up of three or four planes of oxygen held together by planes of aluminium and silicon by way of ionic bonds that together form a single layer of clay. The spatial arrangement of the oxygen atoms determines clay's structure. Half of the weight of clay is oxygen, but on a volume basis oxygen is ninety percent. The layers of clay are sometimes held together through hydrogen bonds, sodium or potassium bridges and as a result will swell less in the presence of water. Clays such as montmorillonite have layers that are loosely attached and will swell greatly when water intervenes between the layers.

In a wider sense clays can be classified as:

1. Layer Crystalline alumino-silica clays: montmorillonite, illite, vermiculite, chlorite, kaolinite.
2. Crystalline Chain carbonate and sulfate minerals: calcite (CaCO3), dolomite (CaMg(CO3)2) and gypsum (CaSO4·2H2O).
3. Amorphous clays: young mixtures of silica (SiO2-OH) and alumina (Al(OH)3) which have not had time to form regular crystals.
4. Sesquioxide clays: old, highly leached clays which result in oxides of iron, aluminium and titanium.

Alumino-silica clays: Alumino-silica clays or aluminosilicate clays are characterized by their regular crystalline or quasi-crystalline structure. Oxygen in ionic bonds with silicon forms a tetrahedral coordination (silicon at the center) which in turn forms sheets of silica. Two sheets of silica are bonded together by a plane of aluminium which forms an octahedral coordination, called alumina, with the oxygens of the silica sheet above and that below it. Hydroxyl ions (OH) sometimes substitute for oxygen. During the clay formation process, Al may substitute for Si in the silica layer, and as much as one fourth of the aluminium Al may be substituted by Zn, Mg or Fe in the alumina layer. The substitution of lower-valence cations for higher-valence cations (isomorphous substitution) gives clay a local negative charge on an oxygen atom that attracts and holds water and positively charged soil cations, some of which are of value for plant growth. Isomorphous substitution occurs during the clay's formation and does not change with time.

- Montmorillonite clay is made of four planes of oxygen with two silicon and one central aluminium plane intervening. The aluminosilicate montmorillonite clay is thus said to have a 2:1 ratio of silicon to aluminium, in short it is called a 2:1 clay mineral. The seven planes together form a single crystal of montmorillonite. The crystals are weakly held together and water may intervene, causing the clay to swell up to ten times its dry volume. It occurs in soils which have had little leaching, hence it is found in arid regions, although it may also occur in humid climates, depending on its mineralogical origin. As the crystals are not bonded face to face, the entire surface is exposed and available for surface reactions, hence it has a high cation exchange capacity (CEC).

- Illite is a 2:1 clay similar in structure to montmorillonite but has potassium bridges between the faces of the clay crystals and the degree of swelling depends on the degree of weathering of potassium-feldspar. The active surface area is reduced due to the potassium bonds. Illite originates from the modification of mica, a primary mineral. It is often found together with montmorillonite and its primary minerals. It has moderate CEC.
- Vermiculite is a mica-based clay similar to illite, but the crystals of clay are held together more loosely by hydrated magnesium and it will swell, but not as much as does montmorillonite. It has very high CEC.
- Chlorite is similar to vermiculite, but the loose bonding by occasional hydrated magnesium, as in vermiculite, is replaced by a hydrated magnesium sheet, that firmly bonds the planes above and below it. It has two planes of silicon, one of aluminium and one of magnesium; hence it is a 2:2 clay. Chlorite does not swell and it has low CEC.
- Kaolinite is a very common, highly weathered clay, and more common than montmorillonite in acid soils. It has one silica and one alumina plane per crystal; hence it is a 1:1 type clay. One plane of silica of montmorillonite is dissolved and is replaced with hydroxyls, which produces strong hydrogen bonds to the oxygen in the next crystal of clay. As a result, kaolinite does not swell in water and has a low specific surface area, and as almost no isomorphous substitution has occurred it has a low CEC. Where rainfall is high, acid soils selectively leach more silica than alumina from the original clays, leaving kaolinite. Even heavier weathering results in sesquioxide clays.

Crystalline chain clays: The carbonate and sulfate clay minerals are much more soluble and hence are found primarily in desert soils where leaching is less active.

Amorphous clays: Amorphous clays are young, and commonly found in recent volcanic ash deposits such as tephra. They are mixtures of alumina and silica which have not formed the ordered crystal shape of alumino-silica clays which time would provide. The majority of their negative charges originates from hydroxyl ions, which can gain or lose a hydrogen ion (H) in response to soil pH, in such way as to buffer the soil pH. They may have either a negative charge provided by the attached hydroxyl ion (OH), which can attract a cation, or lose the hydrogen of the hydroxyl to solution and display a positive charge which can attract anions. As a result, they may display either high CEC in an acid soil solution, or high anion exchange capacity in a basic soil solution.

Sesquioxide Clays

silica-Sesquioxide

Sesquioxide clays are a product of heavy rainfall that has leached most of the silica from alumino-silica clay, leaving the less soluble oxides iron hematite (Fe2O3), iron hydroxide (Fe(OH)3), aluminium hydroxide gibbsite (Al(OH)3), hydrated manganese birnessite (MnO2), as can be observed in most lateritic weathering profiles of tropical soils. It takes hundreds of thousands of years of leaching to create sesquioxide clays. Sesqui is Latin for "one and one-half": there are three parts oxygen to two parts iron or aluminium; hence the ratio is one and one-half (not true for all). They are hydrated and act as either amorphous or crystalline. They are not sticky and do not swell, and soils high in them behave much like sand and can rapidly pass water. They are able to hold large quantities of phosphates, a sorptive process which can at least partly be inhibited in the presence of decomposed (humified) organic matter. Sesquioxides have low CEC but these variable-charge minerals are able to hold anions as well as cations. Such soils range from yellow to red in colour. Such clays tend to hold phosphorus so tightly that it is unavailable for absorption by plants.

Organic Colloids

Humus is one of the two final stages of decomposition of organic matter. It remains in the soil as the organic component of the soil matrix while the other stage, carbon dioxide, is freely liberated in the atmosphere or reacts with calcium to form the soluble calcium bicarbonate. While humus may linger for a thousand years, on the larger scale of the age of the mineral soil components, it is temporary, being finally released as CO2. It is composed of the very stable lignins (30%) and complex sugars (polyuronides, 30%), proteins (30%), waxes, and fats that are resistant to breakdown by microbes and can form complexes with metals, facilitating their downward migration (podzolization). However, although originating for its main part from dead plant organs (wood, bark, foliage, roots), a large part of humus comes from organic compounds excreted by soil organisms (roots, microbes, animals) and from their decomposition upon death. Its chemical assay is 60% carbon, 5% nitrogen, some oxygen and the remainder hydrogen, sulfur, and phosphorus. On a dry weight basis, the CEC of humus is many times greater than that of clay.

Humus plays a major role in the regulation of atmospheric carbon, through carbon sequestration in the soil profile, more especially in deeper horizons with reduced biological activity. Stocking and destocking of soil carbon are under strong climate influence. They are normally balanced through an equilibrium between production and mineralization of organic matter, but the balance is in favour of destocking under present-day climate warming, and more especially in permafrost.

Carbon and Terra Preta

In the extreme environment of high temperatures and the leaching caused by the heavy rain of tropical rain forests, the clay and organic colloids are largely destroyed. The heavy rains wash the alumino-silicate clays from the soil leaving only sesquioxide clays of low CEC. The high temperatures and humidity allow bacteria and fungi to virtually decay any organic matter on the rain-forest floor overnight and much of the nutrients are volatilized or leached from the soil and lost, leaving only a thin root mat lying directly on the mineral soil. However, carbon in the form of finely divided charcoal, also known as black carbon, is far more stable than soil colloids and is capable of performing many of the functions of the soil colloids of sub-tropical soils.

Soil containing substantial quantities of charcoal, of an anthropogenic origin, is called terra preta. In Amazonia it testifies for the agronomic knowledge of past Amerindian civilizations. The pantropical peregrine earthworm Pontoscolex corethrurus has been suspected to contribute to the fine division of charcoal and its mixing to the mineral soil in the frame of present-day slash-and-burn or shifting cultivation still practiced by Amerindian tribes. Research into terra preta is still young but is promising. Fallow periods "on the Amazonian Dark Earths can be as short as 6 months, whereas fallow periods on oxisols are usually 8 to 10 years long" The incorporation of charcoal to agricultural soil for improving water and nutrient retention has been called biochar, being extended to other charred or carbon-rich by-products, and is now increasingly used in sustainable tropical agriculture. Biochar also allows the irreversible sorption of pesticides and other pollutants, a mechanism by which their mobility, and thus their environmental risk, decreases. It has also been argued as a mean of sequestering more carbon in the soil, thereby mitigating the so-called greenhouse effect. However, the use of biochar is limited by the availability of wood or other products of pyrolysis and by risks caused by concomitent deforestation.

Soil Biodiversity

Large numbers of microbes, animals, plants and fungi are living in soil. However, biodiversity in soil is much harder to study as most of this life is invisible, hence estimates about soil biodiversity have been unsatisfactory. A recent study suggested that soil is likely home to 59 ± 15% of the species on Earth. Enchytraeidae (worms) have the greatest percentage of species in soil (98.6%), followed by fungi (90%), plants (85.5%), and termites (Isoptera) (84.2%). Many other groups of animals have substantial fractions of species living in soil, e.g. about 30% of insects, and close to 50% of arachnids. While most vertebrates live above ground (ignoring aquatic species), many species are fossorial, that is, they live in soil, such as most blind snakes.

Soil biology is the study of microbial and faunal activity in the soil. This photo shows the activity of both. Soil biology is the study of microbial and faunal activity and ecology in soil. Soil life, soil biota, soil fauna, or edaphon is a collective term that encompasses all organisms that spend a significant portion of their life cycle within a soil profile, or at the soil-litter interface. These organisms include earthworms, nematodes, protozoa, fungi, bacteria, different arthropods, as well as some reptiles (such as snakes), and species of burrowing mammals like gophers, moles and prairie dogs. Soil biology plays a vital role in determining many soil characteristics. The decomposition of organic matter by soil organisms has an immense influence on soil fertility, plant growth, soil structure, and carbon storage. As a relatively new science, much remains unknown about soil biology and its effect on soil ecosystems.

Overview

The soil is home to a large proportion of the world's biodiversity. The links between soil organisms and soil functions are complex. The interconnectedness and complexity of this soil 'food web' means any appraisal of soil function must necessarily take into account interactions with the living communities that exist within the soil. We know that soil organisms break down organic matter, making nutrients available for uptake by plants and other organisms.

The nutrients stored in the bodies of soil organisms prevent nutrient loss by leaching. Microbial exudates act to maintain soil structure, and earthworms are important in bioturbation. However, we find that we don't understand critical aspects about how these populations function and interact.

The discovery of glomalin in 1995 indicates that we lack the knowledge to correctly answer some of the most basic questions about the biogeochemical cycle in soils. There is much work ahead to gain a better understanding of the ecological role of soil biological components in the biosphere.

In balanced soil, plants grow in an active and steady environment. The mineral content of the soil and its heartiful structure are important for their well-being, but it is the life in the earth that powers its cycles and provides its fertility. Without the activities of soil organisms, organic materials would accumulate and litter the soil surface, and there would be no food for plants. The soil biota includes:

- Megafauna: size range - 20 mm upward, e.g. moles, rabbits, and rodents.
- Macrofauna: size range - 2 to 20 mm, e.g. woodlice, earthworms, beetles, centipedes, slugs, snails, ants, and harvestmen.
- Mesofauna: size range - 100 micrometres to 2 mm, e.g. tardigrades, mites and springtails.
- Microfauna and Microflora: size range - 1 to 100 micrometres, e.g. yeasts, bacteria (commonly actinobacteria), fungi, protozoa, roundworms, and rotifers.

Of these, bacteria and fungi play key roles in maintaining a healthy soil. They act as decomposers that break down organic materials to produce detritus and other breakdown products. Soil detritivores, like earthworms, ingest detritus and decompose it. Saprotrophs, well represented by fungi and bacteria, extract soluble nutrients from delitro. The ants (macrofaunas) help by breaking down in the same way but they also provide the motion part as they move in their armies. Also the rodents, wood-eaters help the soil to be more absorbent.

Scope

Soil biology involves work in the following areas:

- Modelling of biological processes and population dynamics
- Soil biology, physics and chemistry: occurrence of physicochemical parameters and surface properties on biological processes and population behavior
- Population biology and molecular ecology: methodological development and contribution to study microbial and faunal populations; diversity and population dynamics; genetic transfers, influence of environmental factors
- Community ecology and functioning processes: interactions between organisms and mineral or organic compounds; involvement of such interactions in soil pathogenicity; transformation of mineral and organic compounds, cycling of elements; soil structuration

Complementary disciplinary approaches are necessarily utilized which involve molecular biology, genetics, ecophysiology, biogeography, ecology, soil processes, organic matter, nutrient dynamics and landscape ecology.

Bacteria

Bacteria are single-cell organisms and the most numerous denizens of agriculture, with populations ranging from 100 million to 3 billion in a gram. They are capable of very rapid reproduction by binary fission (dividing into two) in favourable conditions. One bacterium is capable of producing 16 million more in just 24 hours. Most soil bacteria live close to plant roots and are often referred to as rhizobacteria. Bacteria live in soil water, including the film of moisture surrounding soil particles, and some are able to swim by means of flagella. The majority of the beneficial soil-dwelling bacteria need oxygen (and are thus termed aerobic bacteria), whilst those that do not require air are referred to as anaerobic, and tend to cause putrefaction of dead organic matter. Aerobic bacteria are most active in a soil that is moist (but not saturated, as this will deprive aerobic bacteria of the air that they require), and neutral soil pH, and where there is plenty of food (carbohydrates and micronutrients from organic matter) available. Hostile conditions will not completely kill bacteria;

rather, the bacteria will stop growing and get into a dormant stage, and those individuals with pro-adaptive mutations may compete better in the new conditions. Some gram-positive bacteria produce spores in order to wait for more favourable circumstances, and gram-negative bacteria get into a "nonculturable" stage. Bacteria are colonized by persistent viral agents (bacteriophages) that determine gene word order in bacterial host.

From the organic gardener's point of view, the important roles that bacteria play are:

The Nitrogen Cycles

Nitrification

Nitrification is a vital part of the nitrogen cycle, wherein certain bacteria (which manufacture their own carbohydrate supply without using the process of photosynthesis) are able to transform nitrogen in the form of ammonium, which is produced by the decomposition of proteins, into nitrates, which are available to growing plants, and once again converted to proteins.

Nitrogen fixation: In another part of the cycle, the process of nitrogen fixation constantly puts additional nitrogen into biological circulation. This is carried out by free-living nitrogen-fixing bacteria in the soil or water such as Azotobacter, or by those that live in close symbiosis with leguminous plants, such as rhizobia. These bacteria form colonies in nodules they create on the roots of peas, beans, and related species. These are able to convert nitrogen from the atmosphere into nitrogen-containing organic substances.

Denitrification: While nitrogen fixation converts nitrogen from the atmosphere into organic compounds, a series of processes called denitrification returns an approximately equal amount of nitrogen to the atmosphere. Denitrifying bacteria tend to be anaerobes, or facultatively anaerobes (can alter between the oxygen dependent and oxygen independent types of metabolisms), including Achromobacter and Pseudomonas. The purification process caused by oxygen-free conditions converts nitrates and nitrites in soil into nitrogen gas or into gaseous compounds such as nitrous oxide or nitric oxide. In excess, denitrification can lead to overall losses of available soil nitrogen and subsequent loss of soil fertility. However, fixed nitrogen may circulate many times between organisms and the soil before denitrification returns it to the atmosphere. The diagram above illustrates the nitrogen cycle.

Actinomycetota: Actinomycetota are critical in the decomposition of organic matter and in humus formation. They specialize in breaking down cellulose and lignin along with the tough chitin found on the exoskeletons of insects. Their presence is responsible for the sweet "earthy" aroma associated with a good healthy soil. They require plenty of air and a pH between 6.0 and 7.5, but are more tolerant of dry conditions than most other bacteria and fungi.

Fungi

A gram of garden soil can contain around one million fungi, such as yeasts and moulds. Fungi have no chlorophyll, and are not able to photosynthesise. They cannot use atmospheric carbon dioxide as a source of carbon, therefore they are chemo-heterotrophic, meaning that, like animals, they require a chemical source of energy rather than being able to use light as an energy source, as well as organic substrates to get carbon for growth and development.

Many fungi are parasitic, often causing disease to their living host plant, although some have beneficial relationships with living plants, as illustrated below. In terms of soil and humus creation, the most important fungi tend to be saprotrophic; that is, they live on dead or decaying organic matter, thus breaking it down and converting it to forms that are available to the higher plants. A succession of fungi species will colonise the dead matter, beginning with those that use sugars and starches, which are succeeded by those that are able to break down cellulose and lignins.

Fungi spread underground by sending long thin threads known as mycelium throughout the soil; these threads can be observed throughout many soils and compost heaps. From the mycelia the fungi is able to throw up its fruiting bodies, the visible part above the soil (e.g., mushrooms, toadstools, and puffballs), which may contain millions of spores. When the fruiting body bursts, these spores are dispersed through the air to settle in fresh environments, and are able to lie dormant for up to years until the right conditions for their activation arise or the right food is made available.

Mycorrhizae: Those fungi that are able to live symbiotically with living plants, creating a relationship that is beneficial to both, are known as Mycorrhizae (from myco meaning fungal and rhiza meaning root). Plant root hairs are invaded by the mycelia of the mycorrhiza, which lives partly in the soil and partly in the root, and may either cover the length of the root hair as a sheath or be concentrated around its tip. The mycorrhiza obtains the carbohydrates that it requires from the root, in return providing the plant with nutrients including nitrogen and moisture. Later the plant roots will also absorb the mycelium into its own tissues.

Beneficial mycorrhizal associations are to be found in many of our edible and flowering crops. Shewell Cooper suggests that these include at least 80% of the brassica and solanum families (including tomatoes and potatoes), as well as the majority of tree species, especially in forest and woodlands. Here the mycorrhizae create a fine underground mesh that extends greatly beyond the limits of the tree's roots, greatly increasing their feeding range and actually causing neighbouring trees to become physically interconnected. The benefits of mycorrhizal relations to their plant partners are not limited to nutrients, but can be essential for plant reproduction.

In situations where little light is able to reach the forest floor, such as the North American pine forests, a young seedling cannot obtain sufficient light to photosynthesise for itself and will not grow properly in a sterile soil. But, if the ground is underlain by a mycorrhizal mat, then the developing seedling will throw down roots that can link with the fungal threads and through them obtain the nutrients it needs, often indirectly obtained from its parents or neighbouring trees.

David Attenborough points out the plant, fungi, animal relationship that creates a "Three way harmonious trio" to be found in forest ecosystems, wherein the plant/fungi symbiosis is enhanced by animals such as the wild boar, deer, mice, or flying squirrel, which feed upon the fungi's fruiting bodies, including truffles, and cause their further spread (Private Life Of Plants, 1995). A greater understanding of the complex relationships that pervade natural systems is one of the major justifications of the organic gardener, in refraining from the use of artificial chemicals and the damage these might cause.

Recent research has shown that arbuscular mycorrhizal fungi produce glomalin, a protein that binds soil particles and stores both carbon and nitrogen. These glomalin-related soil proteins are an important part of soil organic matter.

Invertebrates: Soil fauna affect soil formation and soil organic matter dynamically on many spatiotemporal scales. Earthworms, ants and termites mix the soil as they burrow, significantly affecting soil formation. Earthworms ingest soil particles and organic residues, enhancing the availability of plant nutrients in the material that passes through and out of their bodies. By aerating and stirring the soil, and by increasing the stability of soil aggregates, these organisms help to assure the ready infiltration of water. These organisms in the soil also help improve pH levels.

Ants and termites are often referred to as "Soil engineers" because, when they create their nests, there are several chemical and physical changes made to the soil. Among these changes are increasing the presence of the most essential elements like Carbon, Nitrogen, and phosphorus-elements needed for plant growth. They also can gather soil particles from differing depths of soil and deposit them in other places, leading to the mixing of soil so it is richer with nutrients and other elements.

Vertebrates

Figure: *Gopher sticking out of burrow*

The soil is also important to many mammals. Gophers, moles, prairie dogs, and other burrowing animals rely on this soil for protection and food. The animals even give back to the soil as their burrowing allows more rain, snow and water from ice to enter the soil instead of creating erosion.

Chemistry

The chemistry of a soil determines its ability to supply available plant nutrients and affects its physical properties and the health of its living population. In addition, a soil's chemistry also determines its corrosivity, stability, and ability to absorb pollutants and to filter water. It is the surface chemistry of mineral and organic colloids that determines soil's chemical properties. A colloid is a small, insoluble particle ranging in size from 1 nanometer to 1 micrometer, thus small enough to remain suspended by Brownian motion in a fluid medium without settling. Most soils contain organic colloidal particles called humus as well as the inorganic colloidal particles of clays.

The very high specific surface area of colloids and their net electrical charges give soil its ability to hold and release ions. Negatively charged sites on colloids attract and release cations in what is referred to as cation exchange. Cation-exchange capacity is the amount of exchangeable cations per unit weight of dry soil and is expressed in terms of milliequivalents of positively charged ions per 100 grams of soil (or centimoles of positive charge per kilogram of soil; cmolc/kg). Similarly, positively charged sites on colloids can attract and release anions in the soil, giving the soil anion exchange capacity.

Cation and anion exchange: The cation exchange, that takes place between colloids and soil water, buffers (moderates) soil pH, alters soil structure, and purifies percolating water by adsorbing cations of all types, both useful and harmful.

The negative or positive charges on colloid particles make them able to hold cations or anions, respectively, to their surfaces. The charges result from four sources.

1. Isomorphous substitution occurs in clay during its formation, when lower-valence cations substitute for higher-valence cations in the crystal structure. Substitutions in the outermost layers are more effective than for the innermost layers, as the electric charge strength drops off as the square of the distance. The net result is oxygen atoms with net negative charge and the ability to attract cations.
2. Edge-of-clay oxygen atoms are not in balance ionically as the tetrahedral and octahedral structures are incomplete.
3. Hydroxyls may substitute for oxygens of the silica layers, a process called hydroxylation. When the hydrogens of the clay hydroxyls are ionised into solution, they leave the oxygen with a negative charge (anionic clays).
4. Hydrogens of humus hydroxyl groups may also be ionised into solution, leaving, similarly to clay, an oxygen with a negative charge.

Cations held to the negatively charged colloids resist being washed downward by water and are out of reach of plant roots, thereby preserving the soil fertility in areas of moderate rainfall and low temperatures.

There is a hierarchy in the process of cation exchange on colloids, as cations differ in the strength of adsorption by the colloid and hence their ability to replace one another (ion exchange). If present in equal amounts in the soil water solution:

Al replaces H replaces Ca replaces Mg replaces K same as NH

4 replaces Na: If one cation is added in large amounts, it may replace the others by the sheer force of its numbers. This is called law of mass action. This is largely what occurs with the addition of cationic fertilisers (potash, lime).

As the soil solution becomes more acidic (low pH, meaning an abundance of H), the other cations more weakly bound to colloids are pushed into solution as hydrogen ions occupy exchange sites (protonation). A low pH may cause the hydrogen of hydroxyl groups to be pulled into solution, leaving charged sites on the colloid available to be occupied by other cations.

This ionisation of hydroxy groups on the surface of soil colloids creates what is described as pH-dependent surface charges. Unlike permanent charges developed by isomorphous substitution, pH-dependent charges are variable and increase with increasing pH. Freed cations can be made available to plants but are also prone to be leached from the soil, possibly making the soil less fertile. Plants are able to excrete H into the soil through the synthesis of organic acids and by that means, change the pH of the soil near the root and push cations off the colloids, thus making those available to the plant.

Cation exchange capacity (CEC): Cation exchange capacity is the soil's ability to remove cations from the soil water solution and sequester those to be exchanged later as the plant roots release hydrogen ions to the solution. CEC is the amount of exchangeable hydrogen cation (H) that will combine with 100 grams dry weight of soil and whose measure is one milliequivalents per 100 grams of soil (1 meq/100 g). Hydrogen ions have a single charge and one-thousandth of a gram of hydrogen ions per 100 grams dry soil gives a measure of one milliequivalent of hydrogen ion.

Calcium, with an atomic weight 40 times that of hydrogen and with a valence of two, converts to $(40 \div 2) \times 1$ milliequivalent = 20 milliequivalents of hydrogen ion per 100 grams of dry soil or 20 meq/100 g. The modern measure of CEC is expressed as centimoles of positive charge per kilogram (cmol/kg) of oven-dry soil.

Most of the soil's CEC occurs on clay and humus colloids, and the lack of those in hot, humid, wet climates (such as tropical rainforests), due to leaching and decomposition, respectively, explains the apparent sterility of tropical soils. Live plant roots also have some CEC, linked to their specific surface area.

Cation exchange capacity for soils; soil textures; soil colloids		
Soil	**State**	**CEC meq/100 g**
Charlotte fine sand	Florida	1.0
Ruston fine sandy loam	Texas	1.9
Glouchester loam	New Jersey	11.9
Grundy silt loam	Illinois	26.3
Gleason clay loam	California	31.6
Susquehanna clay loam	Alabama	34.3
Davie mucky fine sand	Florida	100.8
Sands	—	1–5
Fine sandy loams	—	5–10
Loams and silt loams	—	5–15
Clay loams	—	15–30
Clays	—	over 30
Sesquioxides	—	0–3
Kaolinite	—	3–15
Illite	—	25–40
Montmorillonite	—	60–100
Vermiculite (similar to illite)	—	80–150
Humus	—	100–300

Anion exchange capacity (AEC): Anion exchange capacity is the soil's ability to remove anions (such as nitrate, phosphate) from the soil water solution and sequester those for later exchange as the plant roots release carbonate anions to the soil water solution. Those colloids which have low CEC tend to have some AEC. Amorphous and sesquioxide clays have the highest AEC, followed by the iron oxides. Levels of AEC are much lower than for CEC, because of the generally higher rate of positively (versus negatively) charged surfaces on soil colloids, to the exception of variable-charge soils. Phosphates tend to be held at anion exchange sites.Iron and aluminum hydroxide clays are able to exchange their hydroxide anions (OH) for other anions. The order reflecting the strength of anion adhesion is as follows:

H_2PO_4 replaces SO_4 replaces NO_3 replaces Cl: The amount of exchangeable anions is of a magnitude of tenths to a few milliequivalents per 100 g dry soil. As pH rises, there are relatively more hydroxyls, which will displace anions from the colloids and force them into solution and out of storage; hence AEC decreases with increasing pH (alkalinity).

Reactivity (pH): Soil reactivity is expressed in terms of pH and is a measure of the acidity or alkalinity of the soil. More precisely, it is a measure of hydronium concentration in an aqueous solution and ranges in values from 0 to 14 (acidic to basic) but practically speaking for soils, pH ranges from 3.5 to 9.5, as pH values beyond those extremes are toxic to life forms.At 25 °C an aqueous solution that has a pH of 3.5 has 10 moles H3O (hydronium ions) per litre of solution (and also 10 moles per litre OH). A pH of 7, defined as neutral, has 10 moles of hydronium ions per litre of solution and also 10 moles of OH per litre; since the two concentrations are equal, they are said to neutralise each other. A pH of 9.5 has 10 moles hydronium ions per litre of solution (and also 10 moles per litre OH). A pH of 3.5 has one million times more hydronium ions per litre than a solution with pH of 9.5 (9.5 - 3.5 = 6 or 10) and is more acidic.

The effect of pH on a soil is to remove from the soil or to make available certain ions. Soils with high acidity tend to have toxic amounts of aluminium and manganese. As a result of a trade-off between toxicity and requirement most nutrients are better available to plants at moderate pH, although most minerals are more soluble in acid soils. Soil organisms are hindered by high acidity, and most agricultural crops do best with mineral soils of pH 6.5 and organic soils of pH 5.5. Given that at low pH toxic metals (e.g. cadmium, zinc, lead) are positively charged as cations and organic pollutants are in non-ionic form, thus both made more available to organisms, it has been suggested that plants, animals and microbes commonly living in acid soils are pre-adapted to every kind of pollution, whether of natural or human origin.

In high rainfall areas, soils tend to acidify as the basic cations are forced off the soil colloids by the mass action of hydronium ions from usual or unusual rain acidity against those attached to the colloids. High rainfall rates can then wash the nutrients out, leaving the soil inhabited only by those organisms which are particularly efficient to uptake nutrients in very acid conditions, like in tropical rainforests.Once the colloids are saturated with H3O, the addition of any more hydronium ions or aluminum hydroxyl cations drives the pH even lower (more acidic) as the soil has been left with no buffering capacity. In areas of extreme rainfall and high temperatures, the clay and humus may be washed out, further reducing the buffering capacity of the soil. In low rainfall areas, unleached calcium pushes pH to 8.5 and with the addition of exchangeable sodium, soils may reach pH 10.

Beyond a pH of 9, plant growth is reduced. High pH results in low micro-nutrient mobility, but water-soluble chelates of those nutrients can correct the deficit. Sodium can be reduced by the addition of gypsum (calcium sulphate) as calcium adheres to clay more tightly than does sodium causing sodium to be pushed into the soil water solution where it can be washed out by an abundance of water.

Base saturation percentage: There are acid-forming cations (e.g. hydronium, aluminium, iron) and there are base-forming cations (e.g. calcium, magnesium, sodium). The fraction of the negatively-charged soil colloid exchange sites (CEC) that are occupied by base-forming cations is called base saturation. If a soil has a CEC of 20 meq and 5 meq are aluminium and hydronium cations (acid-forming), the remainder of positions on the colloids (20- 5 = 15 meq) are assumed occupied by base-forming cations, so that the base saturation is 15 ÷ 20 × 100% = 75% (the compliment 25% is assumed acid-forming cations). Base saturation is almost in direct proportion to pH (it increases with increasing pH).It is of use in calculating the amount of lime needed to neutralise an acid soil (lime requirement). The amount of lime needed to neutralize a soil must take account of the amount of acid forming ions on the colloids (exchangeable acidity), not just those in the soil water solution (free acidity). The addition of enough lime to neutralize the soil water solution will be insufficient to change the pH, as the acid forming cations stored on the soil colloids will tend to restore the original pH condition as they are pushed off those colloids by the calcium of the added lime.

Buffering: The resistance of soil to change in pH, as a result of the addition of acid or basic material, is a measure of the buffering capacity of a soil and (for a particular soil type) increases as the CEC increases. Hence, pure sand has almost no buffering ability, though soils high in colloids (whether mineral or organic) have high buffering capacity. Buffering occurs by cation exchange and neutralisation. However, colloids are not the only regulators of soil pH. The role of carbonates should be underlined, too. More generally, according to pH levels, several buffer systems take precedence over each other, from calcium carbonate buffer range to iron buffer range.

The addition of a small amount of highly basic aqueous ammonia to a soil will cause the ammonium to displace hydronium ions from the colloids, and the end product is water and colloidally fixed ammonium, but little permanent change overall in soil pH.The addition of a small amount of lime, $Ca(OH)_2$, will displace hydronium ions from the soil colloids, causing the fixation of calcium to colloids and the evolution of CO_2 and water, with little permanent change in soil pH.The above are examples of the buffering of soil pH. The general principal is that an increase in a particular cation in the soil water solution will cause that cation to be fixed to colloids (buffered) and a decrease in solution of that cation will cause it to be withdrawn from the colloid and moved into solution (buffered). The degree of buffering is often related to the CEC of the soil; the greater the CEC, the greater the buffering capacity of the soil.

Redox: Soil chemical reactions involve some combination of proton and electron transfer. Oxidation occurs if there is a loss of electrons in the transfer process while reduction occurs if there is a gain of electrons. Reduction potential is measured in volts or millivolts. Soil microbial communities develop along electron transport chains, forming electrically conductive biofilms, and developing networks of bacterial nanowires.Redox factors in soil development, where formation of redoximorphic colour features provides critical information for soil interpretation. Understanding the redox gradient is important to managing carbon sequestration, bioremediation, wetland delineation, and soil-based microbial fuel cells.

Nutrients

There are 15 essential elements that plants must have in order to grow properly. Since the soil provides most essential nutrients, it is crucial that we understand the soil processes that determine the availability of each essential nutrient for plant uptake.Plant nutrients, their chemical symbols, and the ionic forms common in soils and available for plant uptake

Plant nutrients, their chemical symbols, and the ionic forms common in soils and available for plant uptake		
Element	**Symbol**	**Ion or molecule**
Carbon	C	CO_2 (mostly through leaves)
Hydrogen	H	H, H_2O (water)
Oxygen	O	O, OH, CO_3, SO, CO_2
Phosphorus	P	H_2PO_4, HPO_4 (phosphates)
Potassium	K	K
Nitrogen	N	NH_4, NO_3 (ammonium, nitrate)
Sulfur	S	SO_4
Calcium	Ca	Ca
Iron	Fe	Fe, Fe (ferrous, ferric)
Magnesium	Mg	Mg
Boron	B	H_3BO_3, H_2BO_3, $B(OH)_4$
Manganese	Mn	Mn
Copper	Cu	Cu
Zinc	Zn	Zn
Molybdenum	Mo	MoO_4 (molybdate)
Chlorine	Cl	Cl (chloride)

Seventeen elements or nutrients are essential for plant growth and reproduction. They are carbon (C), hydrogen (H), oxygen (O), nitrogen (N), phosphorus (P), potassium (K), sulfur (S), calcium (Ca), magnesium (Mg), iron (Fe), boron (B), manganese (Mn), copper (Cu), zinc (Zn), molybdenum (Mo), nickel (Ni) and chlorine (Cl).

Nutrients required for plants to complete their life cycle are considered essential nutrients. Nutrients that enhance the growth of plants but are not necessary to complete the plant's life cycle are considered non-essential. With the exception of carbon, hydrogen and oxygen, which are supplied by carbon dioxide and water, and nitrogen, provided through nitrogen fixation, the nutrients derive originally from the mineral component of the soil. The Law of the Minimum expresses that when the available form of a nutrient is not in enough proportion in the soil solution, then other nutrients cannot be taken up at an optimum rate by a plant. A particular nutrient ratio of the soil solution is thus mandatory for optimizing plant growth, a value which might differ from nutrient ratios calculated from plant composition.

Plant uptake of nutrients can only proceed when they are present in a plant-available form. In most situations, nutrients are absorbed in an ionic form from (or together with) soil water. Although minerals are the origin of most nutrients, and the bulk of most nutrient elements in the soil is held in crystalline form within primary and secondary minerals, they weather too slowly to support rapid plant growth. For example, the application of finely ground minerals, feldspar and apatite, to soil seldom provides the necessary amounts of potassium and phosphorus at a rate sufficient for good plant growth, as most of the nutrients remain bound in the crystals of those minerals.

The nutrients adsorbed onto the surfaces of clay colloids and soil organic matter provide a more accessible reservoir of many plant nutrients (e.g. K, Ca, Mg, P, Zn). As plants absorb the nutrients from the soil water, the soluble pool is replenished from the surface-bound pool. The decomposition of soil organic matter by microorganisms is another mechanism whereby the soluble pool of nutrients is replenished - this is important for the supply of plant-available N, S, P, and B from soil.

Gram for gram, the capacity of humus to hold nutrients and water is far greater than that of clay minerals, most of the soil cation exchange capacity arising from charged carboxylic groups on organic matter. However, despite the great capacity of humus to retain water once water-soaked, its high hydrophobicity decreases its wettability once dry. All in all, small amounts of humus may remarkably increase the soil's capacity to promote plant growth.

Soil Organic Matter

The organic material in soil is made up of organic compounds and includes plant, animal and microbial material, both living and dead. A typical soil has a biomass composition of 70% microorganisms, 22% macrofauna, and 8% roots. The living component of an acre of soil may include 900 lb of earthworms, 2400 lb of fungi, 1500 lb of bacteria, 133 lb of protozoa and 890 lb of arthropods and algae.

A few percent of the soil organic matter, with small residence time, consists of the microbial biomass and metabolites of bacteria, molds, and actinomycetes that work to break down the dead organic matter. Were it not for the action of these micro-organisms, the entire carbon dioxide part of the atmosphere would be sequestered as organic matter in the soil. However, in the same time soil microbes contribute to carbon sequestration in the topsoil through the formation of stable humus. In the aim to sequester more carbon in the soil for alleviating the greenhouse effect it would be more efficient in the long-term to stimulate humification than to decrease litter decomposition.

The main part of soil organic matter is a complex assemblage of small organic molecules, collectively called humus or humic substances. The use of these terms, which do not rely on a clear chemical classification, has been considered as obsolete. Other studies showed that the classical notion of molecule is not convenient for humus, which escaped most attempts done over two centuries to resolve it in unit components, but still is chemically distinct from polysaccharides, lignins and proteins.

Most living things in soils, including plants, animals, bacteria, and fungi, are dependent on organic matter for nutrients and/or energy. Soils have organic compounds in varying degrees of decomposition which rate is dependent on temperature, soil moisture, and aeration. Bacteria and fungi feed on the raw organic matter, which are fed upon by protozoa, which in turn are fed upon by nematodes, annelids and arthropods, themselves able to consume and transform raw or humified organic matter. This has been called the soil food web, through which all organic matter is processed as in a digestive system. Organic matter holds soils open, allowing the infiltration of air and water, and may hold as much as twice its weight in water. Many soils, including desert and rocky-gravel soils, have little or no organic matter. Soils that are all organic matter, such as peat (histosols), are infertile. In its earliest stage of decomposition, the original organic material is often called raw organic matter. The final stage of decomposition is called humus.

In grassland, much of the organic matter added to the soil is from the deep, fibrous, grass root systems. By contrast, tree leaves falling on the forest floor are the principal source of soil organic matter in the forest. Another difference is the frequent occurrence in the grasslands of fires that destroy large amounts of aboveground material but stimulate even greater contributions from roots. Also, the much greater acidity under any forests inhibits the action of certain soil organisms that otherwise would mix much of the surface litter into the mineral soil. As a result, the soils under grasslands generally develop a thicker A horizon with a deeper distribution of organic matter than in comparable soils under forests, which characteristically store most of their organic matter in the forest floor (O horizon) and thin A horizon.

Humus: Humus refers to organic matter that has been decomposed by soil microflora and fauna to the point where it is resistant to further breakdown. Humus usually constitutes only five percent of the soil or less by volume, but it is an essential source of nutrients and adds important textural qualities crucial to soil health and plant growth. Humus also feeds arthropods, termites and earthworms which further improve the soil. The end product, humus, is suspended in colloidal form in the soil solution and forms a weak acid that can attack silicate minerals by chelating their iron and aluminum atoms. Humus has a high cation and anion exchange capacity that on a dry weight basis is many times greater than that of clay colloids. It also acts as a buffer, like clay, against changes in pH and soil moisture.

Humic acids and fulvic acids, which begin as raw organic matter, are important constituents of humus. After the death of plants, animals, and microbes, microbes begin to feed on the residues through their production of extra-cellular soil enzymes, resulting finally in the formation of humus. As the residues break down, only molecules made of aliphatic and aromatic hydrocarbons, assembled and stabilized by oxygen and hydrogen bonds, remain in the form of complex molecular assemblages collectively called humus. Humus is never pure in the soil, because it reacts with metals and clays to form complexes which further contribute to its stability and to soil structure. Although the structure of humus has in itself few nutrients (with the exception of constitutive metals such as calcium, iron and aluminum) it is able to attract and link, by weak bonds, cation and anion nutrients that can further be released into the soil solution in response to selective root uptake and changes in soil pH, a process of paramount importance for the maintenance of fertility in tropical soils.

Lignin is resistant to breakdown and accumulates within the soil. It also reacts with proteins, which further increases its resistance to decomposition, including enzymatic decomposition by microbes. Fats and waxes from plant matter have still more resistance to decomposition and persist in soils for thousand years, hence their use as tracers of past vegetation in buried soil layers. Clay soils often have higher organic contents that persist longer than soils without clay as the organic molecules adhere to and are stabilised by the clay. Proteins normally decompose readily, to the exception of scleroproteins, but when bound to clay particles they become more resistant to decomposition. As for other proteins clay particles absorb the enzymes exuded by microbes, decreasing enzyme activity while protecting extracellular enzymes from degradation. The addition of organic matter to clay soils can render that organic matter and any added nutrients inaccessible to plants and microbes for many years. A study showed increased soil fertility following the addition of mature compost to a clay soil. High soil tannin content can cause nitrogen to be sequestered as resistant tannin-protein complexes.

Humus formation is a process dependent on the amount of plant material added each year and the type of base soil. Both are affected by climate and the type of organisms present. Soils with humus can vary in nitrogen content but typically have 3 to 6 percent nitrogen. Raw organic matter, as a reserve of nitrogen and phosphorus, is a vital component affecting soil fertility. Humus also absorbs water, and expands and shrinks between dry and wet states to a higher extent than clay, increasing soil porosity. Humus is less stable than the soil's mineral constituents, as it is reduced by microbial decomposition, and over time its concentration diminishes without the addition of new organic matter. However, humus in its most stable forms may persist over centuries if not millennia. Charcoal is a source of highly stable humus, called black carbon, which had been used traditionally to improve the fertility of nutrient-poor tropical soils. This very ancient practice, as ascertained in the genesis of Amazonian dark earths, has been renewed and became popular under the name of biochar. It has been suggested that biochar could be used to sequester more carbon in the fight against the greenhouse effect.

Climatological influence: The production, accumulation and degradation of organic matter are greatly dependent on climate. For example, when a thawing event occurs, the flux of soil gases with atmospheric gases is significantly influenced. Temperature, soil moisture and topography are the major factors affecting the accumulation of organic matter in soils. Organic matter tends to accumulate under wet or cold conditions where decomposer activity is impeded by low temperature or excess moisture which results in anaerobic conditions. Conversely, excessive rain and high temperatures of tropical climates enables rapid decomposition of organic matter and leaching of plant nutrients. Forest ecosystems on these soils rely on efficient recycling of nutrients and plant matter by the living plant and microbial biomass to maintain their productivity, a process which is disturbed by human activities. Excessive slope, in particular in the presence of cultivation for the sake of agriculture, may encourage the erosion of the top layer of soil which holds most of the raw organic material that would otherwise eventually become humus.

Plant residue: Cellulose and hemicellulose undergo fast decomposition by fungi and bacteria, with a half-life of 12-18 days in a temperate climate. Brown rot fungi can decompose the cellulose and hemicellulose, leaving the lignin and phenolic compounds behind. Starch, which is an energy storage system for plants, undergoes fast decomposition by bacteria and fungi. Lignin consists of polymers composed of 500 to 600 units with a highly branched, amorphous structure, linked to cellulose, hemicellulose and pectin in plant cell walls. Lignin undergoes very slow decomposition, mainly by white rot fungi and actinomycetes; its half-life under temperate conditions is about six months.

Horizons

A horizontal layer of the soil, whose physical features, composition and age are distinct from those above and beneath, is referred to as a soil horizon. The naming of a horizon is based on the type of material of which it is composed. Those materials reflect the duration of specific processes of soil formation. They are labelled using a shorthand notation of letters and numbers which describe the horizon in terms of its colour, size, texture, structure, consistency, root quantity, pH, voids, boundary characteristics and presence of nodules or concretions. No soil profile has all the major horizons. Some, called entisols, may have only one horizon or are currently considered as having no horizon, in particular incipient soils from unreclaimed mining waste deposits, moraines, volcanic cones sand dunes or alluvial terraces. Upper soil horizons may be lacking in truncated soils following wind or water ablation, with concomitant downslope burying of soil horizons, a natural process aggravated by agricultural practices such as tillage. The growth of trees is another source of disturbance, creating a micro-scale heterogeneity which is still visible in soil horizons once trees have died. By passing from a horizon to another, from the top to the bottom of the soil profile, one goes back in time, with past events registered in soil horizons like in sediment layers. Sampling pollen, testate amoebae and plant remains in soil horizons may help to reveal environmental changes (e.g. climate change, land use change) which occurred in the course of soil formation. Soil horizons can be dated by several methods such as radiocarbon, using pieces of charcoal provided they are of enough size to escape pedoturbation by earthworm activity and other mechanical disturbances. Fossil soil horizons from paleosols can be found within sedimentary rock sequences, allowing the study of past environments.

The exposure of parent material to favourable conditions produces mineral soils that are marginally suitable for plant growth, as is the case in eroded soils. The growth of vegetation results in the production of organic residues which fall on the ground as litter for plant aerial parts (leaf litter) or are directly produced belowground for subterranean plant organs (root litter), and then release dissolved organic matter. The remaining surficial organic layer, called the O horizon, produces a more active soil due to the effect of the organisms that live within it. Organisms colonise and break down organic materials, making available nutrients upon which other plants and animals can live. After sufficient time, humus moves downward and is deposited in a distinctive organic-mineral surface layer called the A horizon, in which organic matter is mixed with mineral matter through the activity of burrowing animals, a process called pedoturbation. This natural process does not go to completion in the presence of conditions detrimental to soil life such as strong acidity, cold climate or pollution, stemming in the accumulation of undecomposed organic matter within a single organic horizon overlying the mineral soil and in the juxtaposition of humified organic matter and mineral particles, without intimate mixing, in the underlying mineral horizons.

Classification

One of the first soil classification systems was developed by Russian scientist Vasily Dokuchaev around 1880. It was modified a number of times by American and European researchers and was developed into the system commonly used until the 1960s. It was based on the idea that soils have a particular morphology based on the materials and factors that form them. In the 1960s, a different classification system began to emerge which focused on soil morphology instead of parental materials and soil-forming factors. Since then, it has undergone further modifications. The World Reference Base for Soil Resources aims to establish an international reference base for soil classification.

Uses

Soil is used in agriculture, where it serves as the anchor and primary nutrient base for plants. The types of soil and available moisture determine the species of plants that can be cultivated. Agricultural soil science was the primeval domain of soil knowledge, long time before the advent of pedology in the 19th century. However, as demonstrated by aeroponics, aquaponics and hydroponics, soil material is not an absolute essential for agriculture, and soilless cropping systems have been claimed as the future of agriculture for an endless growing mankind.

Soil material is also a critical component in mining, construction and landscape development industries. Soil serves as a foundation for most construction projects. The movement of massive volumes of soil can be involved in surface mining, road building and dam construction. Earth sheltering is the architectural practice of using soil for external thermal mass against building walls. Many building materials are soil based. Loss of soil through urbanization is growing at a high rate in many areas and can be critical for the maintenance of subsistence agriculture.

Soil resources are critical to the environment, as well as to food and fibre production, producing 98.8% of food consumed by humans. Soil provides minerals and water to plants according to several processes involved in plant nutrition. Soil absorbs rainwater and releases it later, thus preventing floods and drought, flood regulation being one of the major ecosystem services provided by soil. Soil cleans water as it percolates through it. Soil is the habitat for many organisms: the major part of known and unknown biodiversity is in the soil, in the form of earthworms, woodlice, millipedes, centipedes, snails, slugs, mites, springtails, enchytraeids, nematodes, protists), bacteria, archaea, fungi and algae; and most organisms living above ground have part of them (plants) or spend part of their life cycle (insects) below-ground. Above-ground and below-ground biodiversities are tightly interconnected, making soil protection of paramount importance for any restoration or conservation plan.

The biological component of soil is an extremely important carbon sink since about 57% of the biotic content is carbon. Even in deserts, cyanobacteria, lichens and mosses form biological soil crusts which capture and sequester a significant amount of carbon by photosynthesis. Poor farming and grazing methods have degraded soils and released much of this sequestered carbon to the atmosphere. Restoring the world's soils could offset the effect of increases in greenhouse gas emissions and slow global warming, while improving crop yields and reducing water needs.

Waste management often has a soil component. Septic drain fields treat septic tank effluent using aerobic soil processes. Land application of waste water relies on soil biology to aerobically treat BOD. Alternatively, landfills use soil for daily cover, isolating waste deposits from the atmosphere and preventing unpleasant smells. Composting is now widely used to treat aerobically solid domestic waste and dried effluents of settling basins. Although compost is not soil, biological processes taking place during composting are similar to those occurring during decomposition and humification of soil organic matter.

Organic soils, especially peat, serve as a significant fuel and horticultural resource. Peat soils are also commonly used for the sake of agriculture in Nordic countries, because peatland sites, when drained, provide fertile soils for food production. However, wide areas of peat production, such as rain-fed sphagnum bogs, also called blanket bogs or raised bogs, are now protected because of their patrimonial interest. As an example, Flow Country, covering 4,000 square kilometres of rolling expanse of blanket bogs in Scotland, is now candidate for being included in the World Heritage List. Under present-day global warming peat soils are thought to be involved in a self-reinforcing (positive feedback) process of increased emission of greenhouse gases (methane and carbon dioxide) and increased temperature, a contention which is still under debate when replaced at field scale and including stimulated plant growth.

Geophagy is the practice of eating soil-like substances. Both animals and humans occasionally consume soil for medicinal, recreational, or religious purposes. It has been shown that some monkeys consume soil, together with their preferred food (tree foliage and fruits), in order to alleviate tannin toxicity.

Soils filter and purify water and affect its chemistry. Rain water and pooled water from ponds, lakes and rivers percolate through the soil horizons and the upper rock strata, thus becoming groundwater. Pests (viruses) and pollutants, such as persistent organic pollutants (chlorinated pesticides, polychlorinated biphenyls), oils (hydrocarbons), heavy metals (lead, zinc, cadmium), and excess nutrients (nitrates, sulfates, phosphates) are filtered out by the soil. Soil organisms metabolise them or immobilise them in their biomass and necromass, thereby incorporating them into stable humus. The physical integrity of soil is also a prerequisite for avoiding landslides in rugged landscapes.

Degradation

Land degradation is a human-induced or natural process which impairs the capacity of land to function. Soil degradation involves acidification, contamination, desertification, erosion or salination.

Acidification

Soil acidification is beneficial in the case of alkaline soils, but it degrades land when it lowers crop productivity, soil biological activity and increases soil vulnerability to contamination and erosion. Soils are initially acid and remain such when their parent materials are low in basic cations (calcium, magnesium, potassium and sodium). On parent materials richer in weatherable minerals acidification occurs when basic cations are leached from the soil profile by rainfall or exported by the harvesting of forest or agricultural crops. Soil acidification is accelerated by the use of acid-forming nitrogenous fertilizers and by the effects of acid precipitation. Deforestation is another cause of soil acidification, mediated by increased leaching of soil nutrients in the absence of tree canopies.

Contamination

Soil contamination at low levels is often within a soil's capacity to treat and assimilate waste material. Soil biota can treat waste by transforming it, mainly through microbial enzymatic activity. Soil organic matter and soil minerals can adsorb the waste material and decrease its toxicity, although when in colloidal form they may transport the adsorbed contaminants to subsurface environments. Many waste treatment processes rely on this natural bioremediation capacity. Exceeding treatment capacity can damage soil biota and limit soil function. Derelict soils occur where industrial contamination or other development activity damages the soil to such a degree that the land cannot be used safely or productively. Remediation of derelict soil uses principles of geology, physics, chemistry and biology to degrade, attenuate, isolate or remove soil contaminants to restore soil functions and values. Techniques include leaching, air sparging, soil conditioners, phytoremediation, bioremediation and Monitored Natural Attenuation. An example of diffuse pollution with contaminants is copper accumulation in vineyards and orchards to which fungicides are repeatedly applied, even in organic farming.

Microfibres from synthetic textiles are another type of plastic soil contamination, 100% of agricultural soil samples from southwestern China contained plastic particles, 92% of which were microfibres. Sources of microfibres likely included string or twine, as well as irrigation water in which clothes had been washed.

The application of biosolids from sewage sludge and compost can introduce microplastics to soils. This adds to the burden of microplastics from other sources (e.g. the atmosphere). Approximately half the sewage sludge in Europe and North America is applied to agricultural land. In Europe it has been estimated that for every million inhabitants 113 to 770 tonnes of microplastics are added to agricultural soils each year.

Desertification

Figure: *Desertification*

Desertification, an environmental process of ecosystem degradation in arid and semi-arid regions, is often caused by badly adapted human activities such as overgrazing or excess harvesting of firewood. It is a common misconception that drought causes desertification. Droughts are common in arid and semiarid lands. Well-managed lands can recover from drought when the rains return. Soil management tools include maintaining soil nutrient and organic matter levels, reduced tillage and increased cover.

These practices help to control erosion and maintain productivity during periods when moisture is available. Continued land abuse during droughts, however, increases land degradation. Increased population and livestock pressure on marginal lands accelerates desertification. It is now questioned whether present-day climate warming will favour or disfavour desertification, with contradictory reports about predicted rainfall trends associated with increased temperature, and strong discrepancies among regions, even in the same country.

Erosion

Figure: *Erosion Control*

Erosion of soil is caused by water, wind, ice, and movement in response to gravity. More than one kind of erosion can occur simultaneously. Erosion is distinguished from weathering, since erosion also transports eroded soil away from its place of origin (soil in transit may be described as sediment). Erosion is an intrinsic natural process, but in many places it is greatly increased by human activity, especially unsuitable land use practices. These include agricultural activities which leave the soil bare during times of heavy rain or strong winds, overgrazing, deforestation, and improper construction activity. Improved management can limit erosion. Soil conservation techniques which are employed include changes of land use (such as replacing erosion-prone crops with grass or other soil-binding plants), changes to the timing or type of agricultural operations, terrace building, use of erosion-suppressing cover materials (including cover crops and other plants), limiting disturbance during construction, and avoiding construction during erosion-prone periods and in erosion-prone places such as steep slopes. Historically, one of the best examples of large-scale soil erosion due to unsuitable land-use practices is wind erosion (the so-called dust bowl) which ruined American and Canadian prairies during the 1930s, when immigrant farmers, encouraged by the federal government of both countries, settled and converted the original shortgrass prairie to agricultural crops and cattle ranching.

A serious and long-running water erosion problem occurs in China, on the middle reaches of the Yellow River and the upper reaches of the Yangtze River. From the Yellow River, over 1.6 billion tons of sediment flow each year into the ocean. The sediment originates primarily from water erosion (gully erosion) in the Loess Plateau region of northwest China. Soil piping is a particular form of soil erosion that occurs below the soil surface. It causes levee and dam failure, as well as sink hole formation. Turbulent flow removes soil starting at the mouth of the seep flow and the subsoil erosion advances up-gradient. The term sand boil is used to describe the appearance of the discharging end of an active soil pipe.

Salination: Soil salination is the accumulation of free salts to such an extent that it leads to degradation of the agricultural value of soils and vegetation. Consequences include corrosion damage, reduced plant growth, erosion due to loss of plant cover and soil structure, and water quality problems due to sedimentation. Salination occurs due to a combination of natural and human-caused processes. Arid conditions favour salt accumulation. This is especially apparent when soil parent material is saline. Irrigation of arid lands is especially problematic. All irrigation water has some level of salinity. Irrigation, especially when it involves leakage from canals and overirrigation in the field, often raises the underlying water table. Rapid salination occurs when the land surface is within the capillary fringe of saline groundwater. Soil salinity control involves watertable control and flushing with higher levels of applied water in combination with tile drainage or another form of subsurface drainage.

Reclamation: Soils which contain high levels of particular clays with high swelling properties, such as smectites, are often very fertile. For example, the smectite-rich paddy soils of Thailand's Central Plains are among the most productive in the world. However, the overuse of mineral nitrogen fertilizers and pesticides in irrigated intensive rice production has endangered these soils, forcing farmers to implement integrated practices based on Cost Reduction Operating Principles.

Many farmers in tropical areas, however, struggle to retain organic matter and clay in the soils they work. In recent years, for example, productivity has declined and soil erosion has increased in the low-clay soils of northern Thailand, following the abandonment of shifting cultivation for a more permanent land use. Farmers initially responded by adding organic matter and clay from termite mound material, but this was unsustainable in the long-term because of rarefaction of termite mounds. Scientists experimented with adding bentonite, one of the smectite family of clays, to the soil. In field trials, conducted by scientists from the International Water Management Institute (IWMI) in cooperation with Khon Kaen University and local farmers, this had the effect of helping retain water and nutrients. Supplementing the farmer's usual practice with a single application of 200 kilograms per rai (1,300 kg/ha; 1,100 lb/acre) of bentonite resulted in an average yield increase of 73%. Other studies showed that applying bentonite to degraded sandy soils reduced the risk of crop failure during drought years.

In 2008, three years after the initial trials, IWMI scientists conducted a survey among 250 farmers in northeast Thailand, half of whom had applied bentonite to their fields. The average improvement for those using the clay addition was 18% higher than for non-clay users. Using the clay had enabled some farmers to switch to growing vegetables, which need more fertile soil. This helped to increase their income. The researchers estimated that 200 farmers in northeast Thailand and 400 in Cambodia had adopted the use of clays, and that a further 20,000 farmers were introduced to the new technique.

If the soil is too high in clay or salts (e.g. saline sodic soil), adding gypsum, washed river sand and organic matter (e.g.municipal solid waste) will balance the composition.

Adding organic matter, like ramial chipped wood or compost, to soil which is depleted in nutrients and too high in sand will boost its quality and improve production. Special mention must be made of the use of charcoal, and more generally biochar to improve nutrient-poor tropical soils, a process based on the higher fertility of anthropogenic pre-Columbian Amazonian Dark Earths, also called Terra Preta de Índio, due to interesting physical and chemical properties of soil black carbon as a source of stable humus. However, the uncontrolled application of charred waste products of all kinds may endanger soil life and human health.

History of Studies and Research

The history of the study of soil is intimately tied to humans' urgent need to provide food for themselves and forage for their animals. Throughout history, civilizations have prospered or declined as a function of the availability and productivity of their soils.

Studies of soil fertility: The Greek historian Xenophon (450-355 BCE) is credited with being the first to expound upon the merits of green-manuring crops: 'But then whatever weeds are upon the ground, being turned into earth, enrich the soil as much as dung.'

Columella's Of husbandry, circa 60 CE, advocated the use of lime and that clover and alfalfa (green manure) should be turned under, and was used by 15 generations (450 years) under the Roman Empire until its collapse. From the fall of Rome to the French Revolution, knowledge of soil and agriculture was passed on from parent to child and as a result, crop yields were low. During the European Middle Ages, Yahya Ibn al-'Awwam's handbook, with its emphasis on irrigation, guided the people of North Africa, Spain and the Middle East; a translation of this work was finally carried to the southwest of the United States when under Spanish influence. Olivier de Serres, considered the father of French agronomy, was the first to suggest the abandonment of fallowing and its replacement by hay meadows within crop rotations. He also highlighted the importance of soil (the French terroir) in the management of vineyards. His famous book Le Théâtre d'Agriculture et mesnage des champs contributed to the rise of modern, sustainable agriculture and to the collapse of old agricultural practices such as soil amendment for crops by the lifting of forest litter and assarting, which ruined the soils of western Europe during the Middle Ages and even later on according to regions.

Experiments into what made plants grow first led to the idea that the ash left behind when plant matter was burned was the essential element but overlooked the role of nitrogen, which is not left on the ground after combustion, a belief which prevailed until the 19th century. In about 1635, the Flemish chemist Jan Baptist van Helmont thought he had proved water to be the essential element from his famous five years' experiment with a willow tree grown with only the addition of rainwater. His conclusion came from the fact that the increase in the plant's weight had apparently been produced only by the addition of water, with no reduction in the soil's weight. John Woodward (d. 1728) experimented with various types of water ranging from clean to muddy and found muddy water the best, and so he concluded that earthy matter was the essential element. Others concluded it was humus in the soil that passed some essence to the growing plant. Still others held that the vital growth principal was something passed from dead plants or animals to the new plants. At the start of the 18th century, Jethro Tull demonstrated that it was beneficial to cultivate (stir) the soil, but his opinion that the stirring made the fine parts of soil available for plant absorption was erroneous.

As chemistry developed, it was applied to the investigation of soil fertility. The French chemist Antoine Lavoisier showed in about 1778 that plants and animals must combust oxygen internally to live. He was able to deduce that most of the 165-pound (75 kg) weight of van Helmont's willow tree derived from air. It was the French agriculturalist Jean-Baptiste Boussingault who by means of experimentation obtained evidence showing that the main sources of carbon, hydrogen and oxygen for plants were air and water, while nitrogen was taken from soil. Justus von Liebig in his book Organic chemistry in its applications to agriculture and physiology (published 1840), asserted that the chemicals in plants must have come from the soil and air and that to maintain soil fertility, the used minerals must be replaced. Liebig nevertheless believed the nitrogen was supplied from the air. The enrichment of soil with guano by the Incas was rediscovered in 1802, by Alexander von Humboldt. This led to its mining and that of Chilean nitrate and to its application to soil in the United States and Europe after 1840.

The work of Liebig was a revolution for agriculture, and so other investigators started experimentation based on it. In England John Bennet Lawes and Joseph Henry Gilbert worked in the Rothamsted Experimental Station, founded by the former, and (re)discovered that plants took nitrogen from the soil, and that salts needed to be in an available state to be absorbed by plants. Their investigations also produced the superphosphate, consisting in the acid treatment of phosphate rock. This led to the invention and use of salts of potassium (K) and nitrogen (N) as fertilizers. Ammonia generated by the production of coke was recovered and used as fertiliser. Finally, the chemical basis of nutrients delivered to the soil in manure was understood and in the mid-19th century chemical fertilisers were applied. However, the dynamic interaction of soil and its life forms was still not understood.

In 1856, J. Thomas Way discovered that ammonia contained in fertilisers was transformed into nitrates, and twenty years later Robert Warington proved that this transformation was done by living organisms. In 1890 Sergei Winogradsky announced he had found the bacteria responsible for this transformation. It was known that certain legumes could take up nitrogen from the air and fix it to the soil but it took the development of bacteriology towards the end of the 19th century to lead to an understanding of the role played in nitrogen fixation by bacteria. The symbiosis of bacteria and leguminous roots, and the fixation of nitrogen by the bacteria, were simultaneously discovered by the German agronomist Hermann Hellriegel and the Dutch microbiologist Martinus Beijerinck. Crop rotation, mechanisation, chemical and natural fertilisers led to a doubling of wheat yields in western Europe between 1800 and 1900.

Studies of soil formation: The scientists who studied the soil in connection with agricultural practices had considered it mainly as a static substrate. However, soil is the result of evolution from more ancient geological materials, under the action of biotic and abiotic processes. After studies of the improvement of the soil commenced, other researchers began to study soil genesis and as a result also soil types and classifications.

In 1860, while in Mississippi, Eugene W. Hilgard (1833-1916) studied the relationship between rock material, climate, vegetation, and the type of soils that were developed. He realised that the soils were dynamic, and considered the classification of soil types. His work was not continued. At about the same time, Friedrich Albert Fallou was describing soil profiles and relating soil characteristics to their formation as part of his professional work evaluating forest and farm land for the principality of Saxony. His 1857 book, Anfangsgründe der Bodenkunde (First principles of soil science) established modern soil science. Contemporary with Fallou's work, and driven by the same need to accurately assess land for equitable taxation, Vasily Dokuchaev led a team of soil scientists in Russia who conducted an extensive survey of soils, observing that similar basic rocks, climate and vegetation types lead to similar soil layering and types, and established the concepts for soil classifications. Due to language barriers, the work of this team was not communicated to western Europe until 1914 through a publication in German by Konstantin Glinka, a member of the Russian team.

Soil Fertility and Productivity-Concepts and Factors

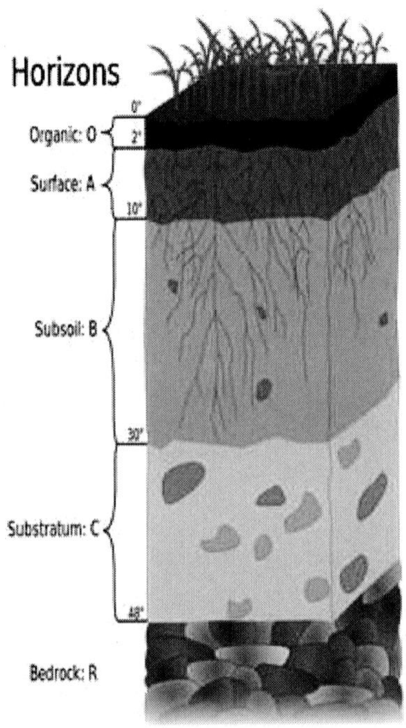

Figure: *Soil scientists use the capital letters O, A, B, C, and E to identify the master horizons, and lowercase letters for distinctions of these horizons. Most soils have three major horizons-the surface horizon (A), the subsoil (B), and the substratum (C). Some soils have an organic horizon (O) on the surface, but this horizon can also be buried. The master horizon, E, is used for subsurface horizons that have a significant loss of minerals (eluviation). Hard bedrock, which is not soil, uses the letter R.*

Soil fertility refers to the ability of soil to sustain agricultural plant growth, i.e. to provide plant habitat and result in sustained and consistent yields of high quality. It also refers to the soil's ability to supply plant/crop nutrients in the right quantities and qualities over a sustained period of time. A fertile soil has the following properties:

- The ability to supply essential plant nutrients and water in adequate amounts and proportions for plant growth and reproduction; and
- The absence of toxic substances which may inhibit plant growth e.g Fe which leads to nutrient toxicity.

The following properties contribute to soil fertility in most situations:

- Sufficient soil depth for adequate root growth and water retention;
- Good internal drainage, allowing sufficient aeration for optimal root growth (although some plants, such as rice, tolerate waterlogging);
- Topsoil or horizon O is with sufficient soil organic matter for healthy soil structure and soil moisture retention;
- Soil pH in the range 5.5 to 7.0 (suitable for most plants but some prefer or tolerate more acid or alkaline conditions);
- Adequate concentrations of essential plant nutrients in plant-available forms;
- Presence of a range of microorganisms that support plant growth.

In lands used for agriculture and other human activities, maintenance of soil fertility typically requires the use of soil conservation practices. This is because soil erosion and other forms of soil degradation generally result in a decline in quality with respect to one or more of the aspects indicated above.

Soil Fertilization

Bioavailable phosphorus (available to soil life) is the element in soil that is most often lacking. Nitrogen and potassium are also needed in substantial amounts. For this reason these three elements are always identified on a commercial fertilizer analysis. For example, a 10-10-15 fertilizer has 10 percent nitrogen, 10 percent available phosphorus (P2O5) and 15 percent water-soluble potassium (K2O). Sulfur is the fourth element that may be identified in a commercial analysis-e.g. 21-0-0-24 which would contain 21% nitrogen and 24% sulfate. Inorganic fertilizers are generally less expensive and have higher concentrations of nutrients than organic fertilizers. Also, since nitrogen, phosphorus and potassium generally must be in the inorganic forms to be taken up by plants, inorganic fertilizers are generally immediately bioavailable to plants without modification. However, some have criticized the use of inorganic fertilizers, claiming that the water-soluble nitrogen doesn't provide for the long-term needs of the plant and creates water pollution. Slow-release fertilizers may reduce leaching loss of nutrients and may make the nutrients that they provide available over a longer period of time.

Soil fertility is a complex process that involves the constant cycling of nutrients between organic and inorganic forms. As plant material and animal wastes are decomposed by micro-organisms, they release inorganic nutrients to the soil solution, a process referred to as mineralization.

Those nutrients may then undergo further transformations which may be aided or enabled by soil micro-organisms. Like plants, many micro-organisms require or preferentially use inorganic forms of nitrogen, phosphorus or potassium and will compete with plants for these nutrients, tying up the nutrients in microbial biomass, a process often called immobilization. The balance between immobilization and mineralization processes depends on the balance and availability of major nutrients and organic carbon to soil microorganisms. Natural processes such as lightning strikes may fix atmospheric nitrogen by converting it to (NO2). Denitrification may occur under anaerobic conditions (flooding) in the presence of denitrifying bacteria. Nutrient cations, including potassium and many micronutrients, are held in relatively strong bonds with the negatively charged portions of the soil in a process known as cation exchange.

In 2008 the cost of phosphorus as fertilizer more than doubled, while the price of rock phosphate as base commodity rose eight-fold. Recently the term peak phosphorus has been coined, due to the limited occurrence of rock phosphate in the world. A wide variety of materials have been described as soil conditioners due to their ability to improve soil quality, including biochar, offering multiple soil health benefits. Food waste compost was found to have better soil improvement than manure based compost.

Light and CO2 Limitations

Photosynthesis is the process whereby plants use light energy to drive chemical reactions which convert CO2 into sugars. As such, all plants require access to both light and carbon dioxide to produce energy, grow and reproduce. While typically limited by nitrogen, phosphorus and potassium, low levels of carbon dioxide can also act as a limiting factor on plant growth. Peer-reviewed and published scientific studies have shown that increasing CO2 is highly effective at promoting plant growth up to levels over 300 ppm. Further increases in co2 can, to a very small degree, continue to increase net photosynthetic output.

Soil Depletion

Soil depletion occurs when the components which contribute to fertility are removed and not replaced, and the conditions which support soil's fertility are not maintained. This leads to poor crop yields. In agriculture, depletion can be due to excessively intense cultivation and inadequate soil management. Soil fertility can be severely challenged when land-use changes rapidly. For example, in Colonial New England, colonists made a number of decisions that depleted the soils, including: allowing herd animals to wander freely, not replenishing soils with manure, and a sequence of events that led to erosion. William Cronon wrote that "...the long-term effect was to put those soils in jeopardy.

The removal of the forest, the increase in destructive floods, the soil compaction and close-cropping wrought by grazing animals, ploughing-all served to increase erosion." One of the most widespread occurrences of soil depletion as of 2008 is in tropical zones where nutrient content of soils is low. The combined effects of growing population densities, large-scale industrial logging, slash-and-burn agriculture and ranching, and other factors, have in some places depleted soils through rapid and almost total nutrient removal.

The depletion of soil has affected the state of plant life and crops in agriculture in many countries. In the Middle East for example, many countries find it difficult to grow produce because of droughts, lack of soil, and lack of irrigation. The Middle East has three countries that indicate a decline in crop production, the highest rates of productivity decline are found in hilly and dryland areas. Many countries in Africa also undergo a depletion of fertile soil. In regions of dry climate like Sudan and the countries that make up the Sahara Desert, droughts and soil degradation is common. Cash crops such as teas, maize, and beans require a variety of nutrients in order to grow healthy. Soil fertility has declined in the farming regions of Africa and the use of artificial and natural fertilizers has been used to regain the nutrients of ground soil.

Topsoil depletion occurs when the nutrient-rich organic topsoil, which takes hundreds to thousands of years to build up under natural conditions, is eroded or depleted of its original organic material. Historically, many past civilizations' collapses can be attributed to the depletion of the topsoil. Since the beginning of agricultural production in the Great Plains of North America in the 1880s, about one-half of its topsoil has disappeared.Depletion may occur through a variety of other effects, including overtillage (which damages soil structure), underuse of nutrient inputs which leads to mining of the soil nutrient bank, and salinization of soil.

Irrigation Effects

The quality of irrigation water is very important to maintain soil fertility and tilth, and for using more soil depth by the plants. When soil is irrigated with high alkaline water, unwanted sodium salts build up in the soil which would make soil draining capacity very poor. So plant roots can not penetrate deep into the soil for optimum growth in Alkali soils.

When soil is irrigated with low pH / acidic water, the useful salts (Ca, Mg, K, P, S, etc.) are removed by draining water from the acidic soil and in addition unwanted aluminium and manganese salts to the plants are dissolved from the soil impeding plant growth. When soil is irrigated with high salinity water or sufficient water is not draining out from the irrigated soil, the soil would convert into saline soil or lose its fertility. Saline water enhance the turgor pressure or osmotic pressure requirement which impedes the off take of water and nutrients by the plant roots.

Top soil loss takes place in alkali soils due to erosion by rain water surface flows or drainage as they form colloids (fine mud) in contact with water. Plants absorb water-soluble inorganic salts only from the soil for their growth. Soil as such does not lose fertility just by growing crops but it lose its fertility due to accumulation of unwanted and depletion of wanted inorganic salts from the soil by improper irrigation and acid rain water (quantity and quality of water). The fertility of many soils which are not suitable for plant growth can be enhanced many times gradually by providing adequate irrigation water of suitable quality and good drainage from the soil.

Arable Land

Arable land ("able to be ploughed") is any land capable of being ploughed and used to grow crops. Alternatively, for the purposes of agricultural statistics, the term often has a more precise definition:

Figure: *Modern mechanised agriculture permits large fields like this one in Dorset, England*

Arable land is the land under temporary agricultural crops (multiple-cropped areas are counted only once), temporary meadows for mowing or pasture, land under market and kitchen gardens and land temporarily fallow (less than five years). The abandoned land resulting from shifting cultivation is not included in this category. Data for 'Arable land' are not meant to indicate the amount of land that is potentially cultivable.

A more concise definition appearing in the Eurostat glossary similarly refers to actual rather than potential uses: "land worked (ploughed or tilled) regularly, generally under a system of crop rotation". In Britain, arable land has traditionally been contrasted with pasturable land such as heaths, which could be used for sheep-rearing but not as farmland.

Arable land is vulnerable to land degradation and some types of un-arable land can be enriched to create useful land. Climate change and biodiversity loss, are driving pressure on arable land.

Plaggen Soil

Plaggen soil or plaggic anthrosol is a type of soil created in parts of northwest Europe in the Middle Ages, as a result of so-called "plaggen" agriculture on marginal podzol soils.

In order to fertilize the fields, pieces of heath or grass including roots and humus ("plaggen") were cut and used as bedding for cattle or sheep. In springtime, this bedding, enriched with slurry was then spread over the fields near the village as manure. The long term practice of this form of agriculture created a rich agricultural soil to a depth of between 40 cm and over 1.50 m, unlike modern arable soils, which tend to be just 30 centimetres deep. The raised fields give rise to a typical landscape with sharp breaks in elevation and are called Plaggenesche in Germany or Es in Dutch. This form of agriculture stopped around 1900 with the introduction of fertilizers.

In Orkney these soils were created already in the 12th to 13th centuries, and on some islands in Shetland these methods continued to be used until the 1960s. The Maori of New Zealand's agriculture included plaggen soil forming practices that increased drainage for kumara crops.

Shifting Cultivation

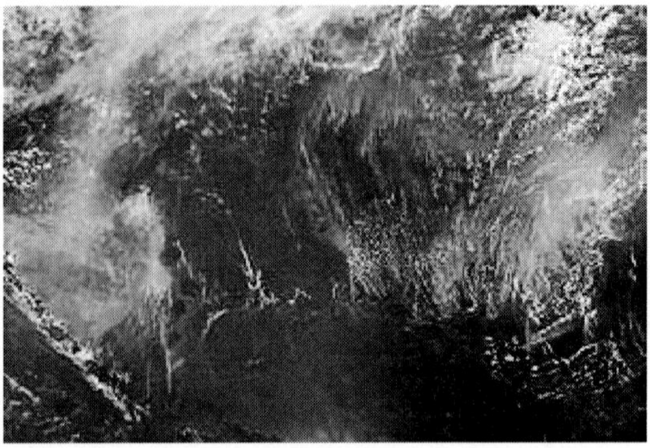

Figure: *Slash-and-burn based shifting cultivation is a widespread historical practice in southeast Asia. Above is a satellite image of Sumatra and Borneo showing shift cultivation fires from October 2006.*

Shifting cultivation is an agricultural system in which plots of land are cultivated temporarily, then abandoned while post-disturbance fallow vegetation is allowed to freely grow while the cultivator moves on to another plot. The period of cultivation is usually terminated when the soil shows signs of exhaustion or, more commonly, when the field is overrun by weeds. The period of time during which the field is cultivated is usually shorter than the period over which the land is allowed to regenerate by lying fallow.

This technique is often used in LEDCs (Less Economically Developed Countries) or LICs (Low Income Countries). In some areas, cultivators use a practice of slash-and-burn as one element of their farming cycle. Others employ land clearing without any burning, and some cultivators are purely migratory and do not use any cyclical method on a given plot. Sometimes no slashing at all is needed where regrowth is purely of grasses, an outcome not uncommon when soils are near exhaustion and need to lie fallow. In shifting agriculture, after two or three years of producing vegetable and grain crops on cleared land, the migrants abandon it for another plot. Land is often cleared by slash-and-burn methods-trees, bushes and forests are cleared by slashing, and the remaining vegetation is burnt. The ashes add potash to the soil. Then the seeds are sown after the rains.

Soil Contamination

Figure: *Excavation showing soil contamination at a disused gasworks in England*

Soil contamination, soil pollution, or land pollution as a part of land degradation is caused by the presence of xenobiotic (human-made) chemicals or other alteration in the natural soil environment. It is typically caused by industrial activity, agricultural chemicals or improper disposal of waste.

The most common chemicals involved are petroleum hydrocarbons, polynuclear aromatic hydrocarbons (such as naphthalene and benzo(a)pyrene), solvents, pesticides, lead, and other heavy metals. Contamination is correlated with the degree of industrialization and intensity of chemical substance.

The concern over soil contamination stems primarily from health risks, from direct contact with the contaminated soil, vapour from the contaminants, or from secondary contamination of water supplies within and underlying the soil. Mapping of contaminated soil sites and the resulting clean ups are time-consuming and expensive tasks, and require expertise in geology, hydrology, chemistry, computer modelling, and GIS in Environmental Contamination, as well as an appreciation of the history of industrial chemistry.

In North America and Western Europe the extent of contaminated land is best known, with many of countries in these areas having a legal framework to identify and deal with this environmental problem. Developing countries tend to be less tightly regulated despite some of them having undergone significant industrialization.

Causes

Soil pollution can be caused by the following (non-exhaustive list):

- Microplastics
- Oil spills
- Mining and activities by other heavy industries
- Accidental spills may happen during activities, etc.
- Corrosion of underground storage tanks (including piping used to transmit the contents)
- Acid rain
- Intensive farming
- Agrochemicals, such as pesticides, herbicides and fertilizers
- Petrochemicals
- Industrial accidents
- Road debris
- Construction activities
- Exterior lead-based paints
- Drainage of contaminated surface water into the soil
- Ammunitions, chemical agents, and other agents of war
- Waste disposal
    - Oil and fuel dumping
    - Nuclear wastes
    - Direct discharge of industrial wastes to the soil
    - Discharge of sewage
    - Landfill and illegal dumping
    - Coal ash
    - Electronic waste
    - Contaminated by rocks containing large amounts of toxic elements.
    - Contaminated by Pb due to vehicle exhaust, Cd, and Zn caused by tire wear.

- o Contamination by strengthening air pollutants by incineration of fossil raw materials.

The most common chemicals involved are petroleum hydrocarbons, solvents, pesticides, lead, and other heavy metals.

Any activity that leads to other forms of soil degradation (erosion, compaction, etc.) may indirectly worsen the contamination effects in that soil remediation becomes more tedious.

Figure: *E-waste processing in Agbogbloshie, Ghana. Improper disposal of manufactured goods and industrial wastes, often means that communities in the global south have to process goods. Especially without proper protections, heavy metals and other contaminates can seep into the soil, and create water pollution and air pollution.*

Historical deposition of coal ash used for residential, commercial, and industrial heating, as well as for industrial processes such as ore smelting, were a common source of contamination in areas that were industrialized before about 1960. Coal naturally concentrates lead and zinc during its formation, as well as other heavy metals to a lesser degree. When the coal is burned, most of these metals become concentrated in the ash (the principal exception being mercury). Coal ash and slag may contain sufficient lead to qualify as a "characteristic hazardous waste", defined in the US as containing more than 5 mg/L of extractable lead using the TCLP procedure.

In addition to lead, coal ash typically contains variable but significant concentrations of polynuclear aromatic hydrocarbons (PAHs; e.g., benzo(a)anthracene, benzo(b)fluoranthene, benzo(k)fluoranthene, benzo(a)pyrene, indeno(cd)pyrene, phenanthrene, anthracene, and others). These PAHs are known human carcinogens and the acceptable concentrations of them in soil are typically around 1 mg/kg. Coal ash and slag can be recognised by the presence of off-white grains in soil, gray heterogeneous soil, or (coal slag) bubbly, vesicular pebble-sized grains.

Treated sewage sludge, known in the industry as biosolids, has become controversial as a "fertilizer". As it is the byproduct of sewage treatment, it generally contains more contaminants such as organisms, pesticides, and heavy metals than other soil.

In the European Union, the Urban Waste Water Treatment Directive allows sewage sludge to be sprayed onto land. The volume is expected to double to 185,000 tons of dry solids in 2005. This has good agricultural properties due to the high nitrogen and phosphate content.

In 1990/1991, 13% wet weight was sprayed onto 0.13% of the land; however, this is expected to rise 15 fold by 2005. Advocates say there is a need to control this so that pathogenic microorganisms do not get into water courses and to ensure that there is no accumulation of heavy metals in the top soil.

Pesticides and Herbicides

A pesticide is a substance used to kill a pest. A pesticide may be a chemical substance, biological agent (such as a virus or bacteria), antimicrobial, disinfectant or device used against any pest. Pests include insects, plant pathogens, weeds, mollusks, birds, mammals, fish, nematodes (roundworms) and microbes that compete with humans for food, destroy property, spread or are a vector for disease or cause a nuisance. Although there are benefits to the use of pesticides, there are also drawbacks, such as potential toxicity to humans and other organisms.

Herbicides are used to kill weeds, especially on pavements and railways. They are similar to auxins and most are biodegradable by soil bacteria.

However, one group derived from trinitrotoluene (2:4 D and 2:4:5 T) have the impurity dioxin, which is very toxic and causes fatality even in low concentrations. Another herbicide is Paraquat. It is highly toxic but it rapidly degrades in soil due to the action of bacteria and does not kill soil fauna.

Insecticides are used to rid farms of pests which damage crops. The insects damage not only standing crops but also stored ones and in the tropics it is reckoned that one third of the total production is lost during food storage. As with fungicides, the first insecticides used in the nineteenth century were inorganic e.g. Paris Green and other compounds of arsenic. Nicotine has also been used since 1690.

There are now two main groups of synthetic insecticides-

1. Organochlorines include DDT, Aldrin, Dieldrin and BHC. They are cheap to produce, potent and persistent. DDT was used on a massive scale from the 1930s, with a peak of 72,000 tonnes used 1970. Then usage fell as the harmful environmental effects were realized. It was found worldwide in fish and birds and was even discovered in the snow in the Antarctic. It is only slightly soluble in water but is very soluble in the bloodstream. It affects the nervous and endocrine systems and causes the eggshells of birds to lack calcium causing them to be easily breakable. It is thought to be responsible for the decline of the numbers of birds of prey like ospreys and peregrine falcons in the 1950s - they are now recovering. As well as increased concentration via the food chain, it is known to enter via permeable membranes, so fish get it through their gills. As it has low water solubility, it tends to stay at the water surface, so organisms that live there are most affected. DDT found in fish that formed part of the human food chain caused concern, but the levels found in the liver, kidney and brain tissues was less than 1 ppm and in fat was 10 ppm, which was below the level likely to cause harm. However, DDT was banned in the UK and the United States to stop the further buildup of it in the food chain. U.S. manufacturers continued to sell DDT to developing countries, who could not afford the expensive replacement chemicals and who did not have such stringent regulations governing the use of pesticides.
2. Organophosphates, e.g. parathion, methyl parathion and about 40 other insecticides are available nationally. Parathion is highly toxic, methyl-parathion is less so and Malathion is generally considered safe as it has low toxicity and is rapidly broken down in the mammalian liver. This group works by preventing normal nerve transmission as cholinesterase is prevented from breaking down the transmitter substance acetylcholine, resulting in uncontrolled muscle movements.

Agents of War

The disposal of munitions, and a lack of care in manufacture of munitions caused by the urgency of production, can contaminate soil for extended periods. There is little published evidence on this type of contamination largely because of restrictions placed by governments of many countries on the publication of material related to war effort.

However, mustard gas stored during World War II has contaminated some sites for up to 50 years and the testing of Anthrax as a potential biological weapon contaminated the whole island of Gruinard.

Soil Biology

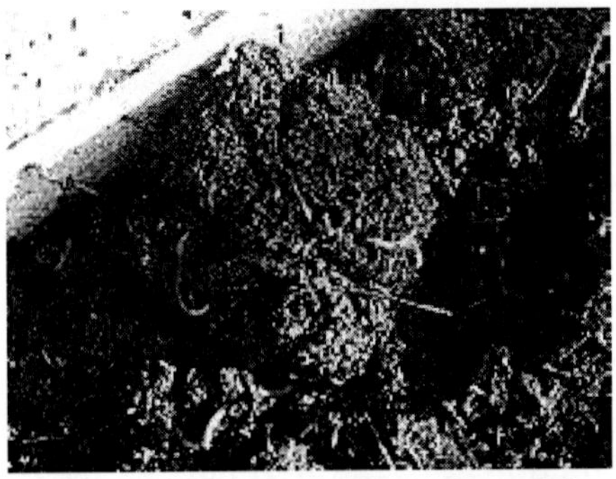

Figure: *Soil biology is the study of microbial and faunal activity in the soil. This photo shows the activity of both.*

Soil biology is the study of microbial and faunal activity and ecology in soil. Soil life, soil biota, soil fauna, or edaphon is a collective term that encompasses all organisms that spend a significant portion of their life cycle within a soil profile, or at the soil-litter interface. These organisms include earthworms, nematodes, protozoa, fungi, bacteria, different arthropods, as well as some reptiles (such as snakes), and species of burrowing mammals like gophers, moles and prairie dogs. Soil biology plays a vital role in determining many soil characteristics. The decomposition of organic matter by soil organisms has an immense influence on soil fertility, plant growth, soil structure, and carbon storage. As a relatively new science, much remains unknown about soil biology and its effect on soil ecosystems.

Overview

The soil is home to a large proportion of the world's biodiversity. The links between soil organisms and soil functions are complex. The interconnectedness and complexity of this soil 'food web' means any appraisal of soil function must necessarily take into account interactions with the living communities that exist within the soil. We know that soil organisms break down organic matter, making nutrients available for uptake by plants and other organisms. The nutrients stored in the bodies of soil organisms prevent nutrient loss by leaching. Microbial exudates act to maintain soil structure, and earthworms are important in bioturbation. However, we find that we don't understand critical aspects about how these populations function and interact. The discovery of glomalin in 1995 indicates that we lack the knowledge to correctly answer some of the most basic questions about the biogeochemical cycle in soils. There is much work ahead to gain a better understanding of the ecological role of soil biological components in the biosphere.

In balanced soil, plants grow in an active and steady environment. The mineral content of the soil and its heartiful structure are important for their well-being, but it is the life in the earth that powers its cycles and provides its fertility. Without the activities of soil organisms, organic materials would accumulate and litter the soil surface, and there would be no food for plants. The soil biota includes:

- Megafauna: size range - 20 mm upward, e.g. moles, rabbits, and rodents.
- Macrofauna: size range - 2 to 20 mm, e.g. woodlice, earthworms, beetles, centipedes, slugs, snails, ants, and harvestmen.
- Mesofauna: size range - 100 micrometres to 2 mm, e.g. tardigrades, mites and springtails.
- Microfauna and Microflora: size range - 1 to 100 micrometres, e.g. yeasts, bacteria (commonly actinobacteria), fungi, protozoa, roundworms, and rotifers.

Of these, bacteria and fungi play key roles in maintaining a healthy soil. They act as decomposers that break down organic materials to produce detritus and other breakdown products. Soil detritivores, like earthworms, ingest detritus and decompose it. Saprotrophs, well represented by fungi and bacteria, extract soluble nutrients from delitro. The ants (macrofaunas) help by breaking down in the same way but they also provide the motion part as they move in their armies. Also the rodents, wood-eaters help the soil to be more absorbent.

Scope

Soil biology involves work in the following areas:

- Modelling of biological processes and population dynamics
- Soil biology, physics and chemistry: occurrence of physicochemical parameters and surface properties on biological processes and population behavior
- Population biology and molecular ecology: methodological development and contribution to study microbial and faunal populations; diversity and population dynamics; genetic transfers, influence of environmental factors

- Community ecology and functioning processes: interactions between organisms and mineral or organic compounds; involvement of such interactions in soil pathogenicity; transformation of mineral and organic compounds, cycling of elements; soil structuration

Complementary disciplinary approaches are necessarily utilized which involve molecular biology, genetics, ecophysiology, biogeography, ecology, soil processes, organic matter, nutrient dynamics and landscape ecology.

Terra Preta

Terra preta (literally "black soil" in Portuguese) is a type of very dark, fertile anthropogenic soil (anthrosol) found in the Amazon Basin. It is also known as "Amazonian dark earth" or "Indian black earth". In Portuguese its full name is terra preta do índio or terra preta de índio ("black soil of the Indian", "Indians' black earth"). Terra mulata ("mulatto earth") is lighter or brownish in colour.

Figure: *Homemade terra preta, with charcoal pieces indicated by white arrows*

Terra preta owes its characteristic black colour to its weathered charcoal content, and was made by adding a mixture of charcoal, bones, broken pottery, compost and manure to the low fertility Amazonian soil. A product of indigenous Amazonian soil management and slash-and-char agriculture, the charcoal is stable and remains in the soil for thousands of years, binding and retaining minerals and nutrients.

Terra preta is characterized by the presence of low-temperature charcoal residues in high concentrations; of high quantities of tiny pottery shards; of organic matter such as plant residues, animal feces, fish and animal bones, and other material; and of nutrients such as nitrogen, phosphorus, calcium, zinc and manganese. Fertile soils such as terra preta show high levels of microorganic activities and other specific characteristics within particular ecosystems.

Terra preta zones are generally surrounded by terra comum or "common soil"; these are infertile soils, mainly acrisols, but also ferralsols and arenosols. Deforested arable soils in the Amazon are productive for a short period of time before their nutrients are consumed or leached away by rain or flooding. This forces farmers to migrate to an unburned area and clear it (by fire). Terra preta is less prone to nutrient leaching because of its high concentration of charcoal, microbial life and organic matter. The combination accumulates nutrients, minerals and microorganisms and withstands leaching.

Terra preta soils were created by farming communities between 450 BCE and 950 CE. Soil depths can reach 2 meters (6.6 ft). It is reported to regenerate itself at the rate of 1 centimeter (0.4 in) per year.

Organic Matter and Nutrients

Charcoal's porosity brings better retention of organic matter, of water and of dissolved nutrients, as well as of pollutants such as pesticides and aromatic poly-cyclic hydrocarbons.

Organic Matter

Charcoal's high absorption potential of organic molecules (and of water) is due to its porous structure. Terra preta's high concentration of charcoal supports a high concentration of organic matter (on average three times more than in the surrounding poor soils), up to 150 g/kg. Organic matter can be found at 1 to 2 metres (3 ft 3 in to 6 ft 7 in) deep.

Bechtold proposes to use terra preta for soils that show, at 50 centimeters (20 in) depth, a minimum proportion of organic matter over 2.0-2.5%. The accumulation of organic matter in moist tropical soils is a paradox, because of optimum conditions for organic matter degradation. It is remarkable that anthrosols regenerate in spite of these tropical conditions' prevalence and their fast mineralisation rates. The stability of organic matter is mainly because the biomass is only partially consumed.

Nutrients

Terra preta soils also show higher quantities of nutrients, and a better retention of these nutrients, than surrounding infertile soils.

The proportion of P reaches 200-400 mg/kg. The quantity of N is also higher in anthrosol, but that nutrient is immobilized because of the high proportion of C over N in the soil.

Anthrosol's availability of P, Ca, Mn and Zn is higher than ferrasol. The absorption of P, K, Ca, Zn, and Cu by the plants increases when the quantity of available charcoal increases. The production of biomass for two crops (rice and Vigna unguiculata) increased by 38-45% without fertilization ($P < 0.05$), compared to crops on fertilized ferralsol.

Amending with charcoal pieces approximately 20 millimeters (0.79 in) in diameter, instead of ground charcoal, did not change the results except for manganese (Mn), for which absorption considerably increased.

Nutrient leaching is minimal in this anthrosol, despite their abundance, resulting in high fertility. When inorganic nutrients are applied to the soil; however, the nutrients' drainage in anthrosol exceeds that in fertilized ferralsol.

As potential sources of nutrients, only C (via photosynthesis) and N (from biological fixation) can be produced in situ. All the other elements (P, K, Ca, Mg, etc.) must be present in the soil. In Amazonia, the provisioning of nutrients from the decomposition of naturally available organic matter fails as the heavy rainfalls wash away the released nutrients and the natural soils (ferralsols, acrisols, lixisols, arenosols, uxisols, etc.) lack the mineral matter to provide those nutrients.

The clay matter that exists in those soils is capable of holding only a small fraction of the nutrients made available from decomposition. In the case of terra preta, the only possible nutrient sources are primary and secondary. The following components have been found:

- Human and animal excrements (rich in P and N);
- Kitchen refuse, such as animal bones and tortoise shells (rich in P and Ca);
- Ash residue from incomplete combustion (rich in Ca, Mg, K, P and charcoal);
- Biomass of terrestrial plants (e.g. compost); and
- Biomass of aquatic plants (e.g. algae).

Saturation in pH and in base is more important than in the surrounding soils.

Microorganisms and Animals

The peregrine earthworm Pontoscolex corethrurus (Oligochaeta: Glossoscolecidae) ingests charcoal and mixes it into a finely ground form with the mineral soil. P. corethrurus is widespread in Amazonia and notably in clearings after burning processes thanks to its tolerance of a low content of organic matter in the soil. This as an essential element in the generation of terra preta, associated with agronomic knowledge involving layering the charcoal in thin regular layers favourable to its burying by P. corethrurus.

Some ants are repelled from fresh terra preta; their density is found to be low about 10 days after production compared to that in control soils.

Modern Research on Creating Terra Preta

Synthetic Terra Preta

A newly coined term is 'synthetic terra preta'. STP is a fertilizer consisting of materials thought to replicate the original materials, including crushed clay, blood and bone meal, manure and biochar is of particulate nature and capable of moving down the soil profile and improving soil fertility and carbon in the current soil peds and aggregates over a viable time frame. Such a mixture provides multiple soil improvements reaching at least the quality of terra mulata. Blood, bone meal and chicken manure are useful for short term organic manure addition. Perhaps the most important and unique part of the improvement of soil fertility is carbon, thought to have been gradually incorporated 4 to 10 thousand years ago. Biochar is capable of decreasing soil acidity and if soaked in nutrient rich liquid can slowly release nutrients and provide habitat for microbes in soil due to its high porosity surface area.

The goal is an economically viable process that could be included in modern agriculture. Average poor tropical soils are easily enrichable to terra preta nova by the addition of charcoal and condensed smoke. Terra preta may be an important avenue of future carbon sequestration while reversing the current worldwide decline in soil fertility and associated desertification. Whether this is possible on a larger scale has yet to be proven. Tree Lucerne (tagasaste or Cytisus proliferus) is one type of fertilizer tree used to make terra preta. Efforts to recreate these soils are underway by companies such as Embrapa and other organizations in Brazil.

Synthetic terra preta is produced at the Sachamama Center for Biocultural Regeneration in High Amazon, Peru. This area has many terra preta soil zones, demonstrating that this anthrosol was created not only in the Amazon basin, but also at higher elevations.

A synthetic terra preta process was developed by Alfons-Eduard Krieger to produce a high humus, nutrient-rich, water-adsorbing soil.

Terra preta Sanitation

Terra preta sanitation (TPS) systems have been studied as an alternative sanitation option by using the effects of lactic-aid conditions in urine-diverting dry toilets and a subsequent treatment by vermicomposting.

Chapter 2

Cation Exchange Capacity

What are Cation Exchange and Cation Exchange Capacity

Soil clay minerals and organic matter tend to be negatively charged, thus attracting positively charged ions (cations) on their surfaces by electrostatic forces. As a result, the cations remain within the soil root zone and are not easily lost through leaching. The adsorbed cations may easily exchange with other cations in the soil solution, hence the term "cation exchange." The adsorbed cations replenish the ions in the soil solution when concentrations decrease due to uptake by plant roots.

Cation exchange capacity (CEC) is a measure of the total negative charges within the soil that adsorb plant nutrient cations such as calcium (Ca), magnesium (Mg) and potassium (K). As such, the CEC is a property of a soil that describes its capacity to supply nutrient cations to the soil solution for plant uptake. Figure 1 illustrates cations retained on soil clay minerals that can exchange with those in the soil solution. Plant roots can remove nutrients from the soil solution, which results in nutrients moving away from the clay particles. Addition of fertilizer to soil causes an initial increase in nutrient concentration in the soil solution, which results in nutrients moving toward clay particles.

The nutrient cations plants use in the largest amounts are potassium (K), calcium (Ca) and magnesium (Mg). Other cations adsorbed on exchange sites are ammonium (NH4), sodium (Na), hydrogen (H), aluminum (Al), iron (Fe or Fe), manganese (Mn), copper (Cu) and zinc (Zn). Micronutrient cations such as zinc, copper, iron and manganese are typically present at very low concentrations in soils. Ammonium concentrations are also typically very low because microorganisms convert ammonium to nitrate in a process called nitrification.

How CEC Changes with Soil PH

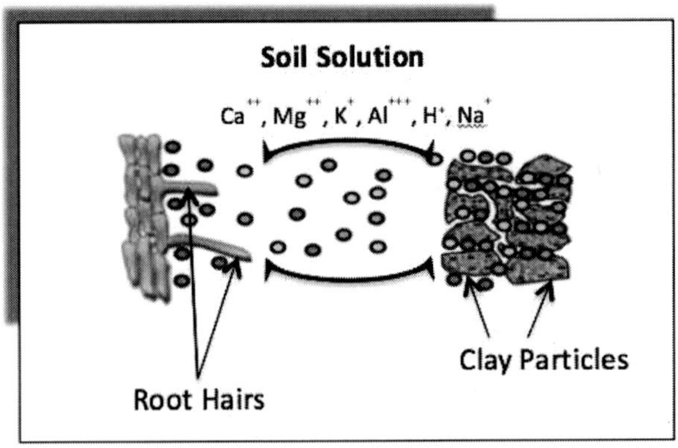

Figure : *Schematic diagram showing exchange of cations between the soil surfaces and the soil solution, and the movement of these cations from soil solution to roots (rhizosphere) for uptake.*

The CEC of soil organic matter and some clay minerals varies with pH. Generally, the CEC is lowest at soil pHs of 3.5 to 4.0 and increases as the pH is increased by liming an acid soil, as shown in Figure 2. Because CEC may vary considerably with soil pH, it is a common practice to measure a soil's CEC at a pH of 7.0. Also note that some positive charges may occur on specific soil mineral surfaces at low pH. These positive charges retain anions (negatively charged ions) such as chloride (Cl) and sulfate (SO4).

Calculating the Cation Exchange Capacity from a Routine Soil Test The CEC value included on typical soil testing laboratory reports is calculated by adding together the concentrations (expressed as milliequivalents of charge per 100 grams of soil) of potassium, magnesium, calcium, sodium and hydrogen, which are extracted from soils using an appropriate extraction method. The University of Georgia Soil Testing Laboratory uses the Mehlich I procedure, based on a double acid (0.05 N HCl + 0.025 N H2SO4) extracting solution. This method is appropriate for acidic, low CEC soils, which are commonly found in Georgia. The CEC of soils containing large amounts of clay or organic matter, or that are alkaline, cannot be satisfactorily analyzed using the Mehlich I extract. Other soil extraction methods should be used on these types of soils.

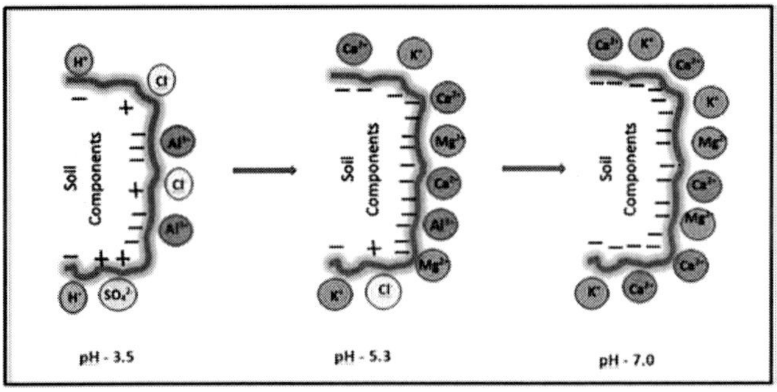

***Figure:** Influence of pH on the surface charge of soil and its components.*

Typical CEC Values in Soils

In most soil reports, CEC is expressed as milliequivalents (meq) of charge (number of charges) per 100 grams of soil (meq/100 g or as cmol/kg when using International Scientific Units). The number of milliequivalents is used rather than a weight (pounds, grams, etc.) of adsorbed cations because CEC represents the total number of charges, which is a better standard of comparison of different soils because each cation species has a different weight and soils differ in the proportions of the different cation species. Table 1 shows the typical cation exchange capacities of soil clay minerals and soils of various textures. Because soil is a mixture of different particle sizes (sand, silt and clay), clay mineral types and organic matter in various proportions, the dominant components and soil pH dictates the soil-s CEC.

Table 1. Cation exchange capacities at pH 7.0 of different soil types, textures and soil organic matter.

Soil and Soil Components	CEC (meq/100 g)
Clay Type	
Kaolinite	3-15
Illite	15-40
Montmorillonite	80-100
Soil Texture	
Sand	1-5
Fine Sandy Loam	5-10
Loam	5-15
Clay Loam	15-30
Clay	>30
Organic Matter	200-400

CEC and Fertility Characteristics of Georgia Soils: For practical purposes, Georgia soils are grouped into four major categories: (1) Coastal Plain, (2) Piedmont, (3) Mountain and Limestone Valley and (4) soils from landscapes, golf greens, greenhouses and flower beds. These categories make for easier evaluation of fertility. The chart below describes the CEC and general fertility characteristics of each soil group (Soil Test Handbook for Georgia, 2008).

Soil Group	**CEC and Soil Fertility Characteristics**
Coastal Plain (includes Atlantic Flatwoods and Sand Hills)	Soils have sandy surfaces and a CEC of 6 meq/100 g or less. Soils in their native state can be acid and infertile. Soils will vary in clay content, drainage characteristics and color. Soils will vary in productivity, ease of handling and adaptation to row crop production. Typical soil types are Norfolk, Lakeland, Lynchburg and Tifton.
Piedmont soils	Soils are predominately upland, well-drained red soils with a CEC of 6 to 12 meq/100 g. Soils in their native state are acid and low in phosphorus but higher in potassium than the Coastal Plain soils. Major soil series are Cecil, Madison and Davison.
Mountain and Limestone Valley soils	Soils may have a gray, sandy surface underlain with a heavy red sandy clay or clay texture soil. The alluvial terraces and river bottoms are gray to light brown in color with yellow to dark red sandy clay loam subsoil. Soils are acid and low in fertility. The average CEC value of these soils is 9 meq/100 g. The major soil types are Porters, Hayesville, Talladega, Fannin, Congaree, Clarkesville, Fullerton, Dewey and Decatur.
Soils from landscapes and golf greens	These soils are frequently maintained differently than crop soils and are usually artificially constituted and maintained. Many are erratic in fertility and cannot be easily placed in one of the three categories given above.

Percent Base Saturation

Percent base saturation (BS) is the percentage of the CEC occupied by the basic cations Ca, Mg and K. Basic cations are distinguished from the acid cations H and Al. At an approximate soil pH 5.4 or less, Al is present in a significantly high concentration that hinders growth of most plant species, and the lower the soil pH, the greater the amount of toxic Al. Therefore, soils with a high percent base saturation are generally more fertile because:

1. They have little or no acid cation Al that is toxic to plant growth.
2. Soils with high percent base saturation have a higher pH; therefore, they are more buffered against acid cations from

plant roots and soil processes that acidify the soil (nitrification, acid rain, etc.).

3. They contain greater amounts of the essential plant nutrient cations K, Ca and Mg for use by plants.

The percentage base saturation is expressed as follows:

$$\%BS = [(Ca + Mg + K)/CEC] \times 100$$

Depending on soil pH, the soil's base saturation may be a fraction of CEC or approximately equal to CEC. In general, if the soil pH is below 7, the base saturation is less than CEC. At pH 7 or higher, soil clay mineral and organic matter surfaces are occupied by basic cations, and thus, base saturation is equal to CEC. Figure 2 illustrates the relative amount of cations retained on soil surfaces at various soil pH levels.

Significance of CEC and BS: A soil's CEC affects fertilization and liming practices. For example, soils with high CEC retain more nutrients than low-CEC soils. With large quantities of fertilizers applied in a single application to sandy soils with low CEC, loss of nutrients is more likely to occur via leaching. In contrast, these nutrients are much less susceptible to losses in clay soils.

Crop production releases acidity into soil. Soil pH will decrease more due to crop production on low CEC soils. High CEC soils are generally well buffered such that pH changes much less from crop production. Therefore, sandy soils low in CEC need to be limed more frequently but at lower rates of application than clay soils. Higher lime rates are needed to reach an optimum pH on high CEC soils due to their greater abundance of acidic cations at a given pH.

Nutrient Availability and pH

The optimum pH for a plant varies with organic matter content and plant type. Plant nutrient availability is strongly tied to the pH in the soil solution. Decreasing soil pH directly increases the solubility of the plant nutrients manganese (Mn), zinc (Zn), copper (Cu), and iron (Fe). Acidic soils make these nutrients more available. At pH values less than about 5.5, toxic levels of Mn, Zn, or aluminum (Al), a non-nutrient element very common in our southern soils, may be released. The impact of pH on nutrient availability is very important-both for maximum plant availability and to avoid potentially toxic levels at very low or very high pH.

The optimal pH for growth differs among plants. For example, regardless of organic matter content, azaleas and blueberries are well-suited for a soil pH of about 5.0. In contrast, asparagus can tolerate a basic soil with a pH up to 8.0. A soil pH of 6.5 to 7.0 is often considered "ideal" for most plants, but a little research can help you identify the proper pH for the plants you wish to grow. After obtaining a soil test report, you can take measures to adjust soil pH or select plants that will thrive at the current pH. Extreme pH measures of 4.0 (acidic) or 10.0 (basic) will support little plant life and are very difficult to modify.

Adjusting Ph

If the soil pH is too basic for the desired plant, incorporating an acidic soil amendment such as pine bark or compost, or applying elemental sulfur, will lower soil pH. Apply sulfur with caution; too much can harm plants.

If the soil pH is too acidic, apply lime to raise the soil pH. There are two general classes of liming materials: calcitic (without magnesium) and dolomitic (with magnesium). Calcitic lime is composed of calcium carbonate (CaCO3) and can be used on soils high in magnesium. Dolomitic lime is a mixture of calcium and magnesium carbonates (CaCO3 and MgCO3), which is the preferred liming material for soils low in magnesium.

Knowing the soil type or even the current pH is not enough to determine the amount of lime needed. The texture of the soil, organic matter content, crop to be grown, target pH, level of soil acidity, CEC, type and amount of clay, and the current pH are all factors to consider in adjusting pH. Soils low in organic matter or high in sand content require less lime to change the pH than clay soils or those with high organic matter.

Lime is heavily regulated in North Carolina. Lime must be labeled with a guarantee of percent calcium and magnesium. The percent of calcium carbonate equivalent also must be included on the label, as well as the pounds of material that equal 1 ton of standard lime.

Each type of lime must meet a screening requirement for particle size. Lime pellets are formed from lime that has been finely ground. The pelleted product is less dusty and easier to apply, but is slower to react with the soil.

Lime moves slowly in the soil and neutralizes acidity only in the area where it is applied. To be effective, it should be spread and thoroughly incorporated. It takes several months for lime to react in the soil, which is why it is good to soil test and plan for proper soil pH management.For established lawns, gardens, and ornamentals that require lime, apply the recommended amount up to 50 pounds of lime per 1,000 square feet in one application to the soil's surface. For recommended rates over 50 pounds, wait several months to make a repeat application to avoid a surface buildup of lime. For new plantings where the area will be tilled, apply the entire recommended amount at one time.

Figure: *Collect soil samples using plastic buckets, let soil air dry and screen out any big chunks (rocks, sticks). Mix several soil samples from the same location before sending it in for testing.*

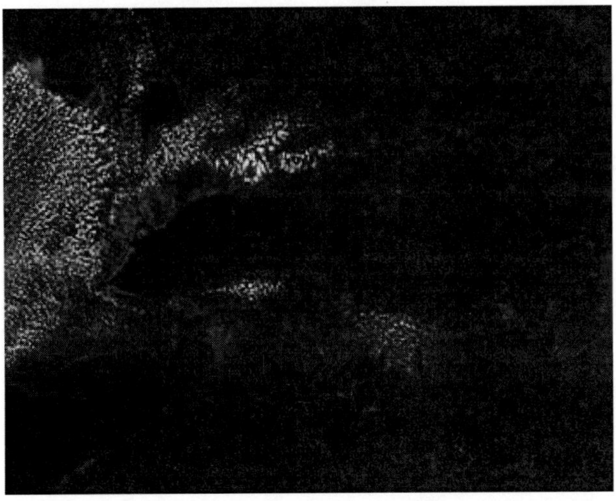

Figure: *The milky blue-green swirls are a phytoplankton bloom in Lake Ontario as seen from space. This bloom is, in part, from nutrient loading (nitrogen and phosphorus) and runoff.*

Figure: *A drop spreader has a cylindrical spinning mechanism (bottom right) that releases fertilizer directly below the spreader as it is pushed along.*

Figure: *A rotary spreader. Bottom left shows the orange disk that spins when the wheels turn. Bottom right shows the spinning mechanism inside the spreader that moves the fertilizer onto the plate below.*

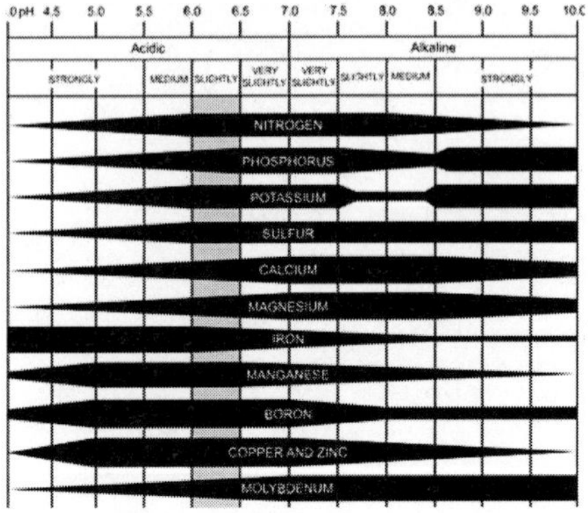

Figure : Nutrient availability as affected by soil pH. The wider areas represent greater availability. The blue bar shows the optimum pH level for nutrient uptake by plants.

Plant Nutrition and Fertilization

Many people confuse plant nutrition with fertilization. Plant nutrition refers to the needs of the plant and how a plant uses the basic chemical elements. Fertilization is the term used when these elements are supplied to the soil as amendments. Adding fertilizer during unfavourable growing conditions will not enhance plant growth and may actually harm or kill plants.

To complete their life cycle, plants need 17 essential nutrients, each in varying amounts. Of these nutrients, three are found in air and water: carbon (C), hydrogen (H), and oxygen (O). Combined, C, H, and O account for about 94% of a plant's weight. The other 6% of a plant's weight includes the remaining 14 nutrients, all of which must come from the soil. Of these, nitrogen (N), phosphorus (P), and potassium (K), the primary macronutrients, are the most needed. Magnesium (Mg), calcium (Ca), and sulfur (S), the secondary macronutrients, are next in the amount needed. The eight other elements-boron, chlorine, copper, iron, manganese, molybdenum, nickel, and zinc-are called micronutrients because they are needed in much smaller amounts than the macronutrients.

Soil Nutrients: For a plant to absorb an element, it must be in a chemical form used by the plant and dissolved in the soil water. In addition to those nutrients already dissolved in soil water, nutrients can be present in the soil in these forms:

- Undissolved or granular form, as from newly applied fertilizer
- Chemicals bound to soil particles
- The chemical structure of soil organic matter released by microbial decomposition

Undissolved or granular nutrients, and those that are chemically bound to soil particles, are not immediately useful, although they have the potential to benefit the plant. For many plant nutrients, the soil acts as a bank. Withdrawals are made from the soil solution, much as you would withdraw money from a checking account. The undissolved pool of soil nutrients is like a savings account. When checking funds are low, transfers are made from the savings account to the checking account. When a checking account is flush with money, some can be moved to savings for long-term retention. In the same way, for many plant nutrients, when the soil solution has excess nutrients, some bind to the soil to become temporarily unavailable, and some react with other chemical elements to form insoluble minerals, which can dissolve again later.

Table 1-3. Relative amounts (out of 100) of the essential nutrients required by most plants.

Primary Nutrients	Amount
Carbon (C)	45
Oxygen (O)	45
Hydrogen (H)	6
Nitrogen (N)	1.5
Potassium (K)	1
Phosphorus (P)	0.2
Secondary Nutrients	**Amount**
Calcium (Ca)	0.5
Magnesium (Mg)	0.2
Sulfur (S)	0.1

Micronutrients	Amount
Iron (Fe)	0.01
Chlorine (Cl)	0.01
Manganese (Mn)	0.005
Boron (B)	0.002
Zinc (Zn)	0.002
Copper (Cu)	0.0006
Molybdenum (Mo)	0.00001

Amounts unknown for Nickel (Ni) and Cobalt (Co)

Several factors improve a plant's ability to use nutrients:

- Type of soil: The more clay and organic matter a soil has, the higher its CEC will be, and the more cationic (positively charged) nutrients it will retain.
- Soil pH: The pH affects how tightly nutrients are bound to soil particles. If the soil pH is extremely high (basic) or very low (acidic), many nutrients become inaccessible to the plant because they are no longer dissolved in the soil water.
- Types of nutrients in the soil: Some nutrients affect the availability of other nutrients. In fact, an apparent deficiency of one nutrient may actually be caused by a large amount of another.
- Amount of soil water: Too much rain leaches nutrients from the soil. If there is too little water, the nutrients cannot dissolve and move into the plant.
- Anything that affects the plant's growth: If growing conditions are good, a plant will absorb nutrients from the soil. If the

plant experiences extremes in temperature, incorrect light levels, or waterlogged or compacted soil, it will have a limited ability to absorb nutrients. Also, plants in dormant stages absorb few nutrients.

The presence or absence of nutrients can cause outward symptoms to appear on the plant. Table 1-4 reviews the essential nutrients for plant growth and symptoms that may appear if a plant is suffering a deficiency or an excess of that nutrient.

Fertilizers

Fertilizers provide some elements that might be lacking in the soil and stimulate healthy, vigorous growth. How much and when to apply fertilizers should be based on observing plant performance, a reliable soil test, and an understanding of the factors that affect growth: light, water, temperature, pests, and nutrition. Simply applying fertilizer because a plant is not growing adequately will not solve many plant problems (insects, disease, or poor drainage, for example), and, in fact, excess nitrogen can often increase insect and disease infestation.

All fertilizers are labeled with three numbers, giving the percentage (by weight) of nitrogen (N), phosphorus (P), and potassium (K). This is referred to as the fertilizer grade.

A 100-pound bag of fertilizer labeled 0-20-10 has 0 pounds of N, 20 pounds of P (reported as P2O5), 10 pounds of K (reported as K2O), and 70 pounds of filler. Filler is added to make the fertilizer easier to spread and to reduce the likelihood of burning plants with too much fertilizer (the fertilizer salts can pull water out of the plant). A fertilizer may also contain secondary macronutrients or micronutrients not listed on the label because the manufacturer does not want to guarantee their exact amounts. Fertilizers can be divided into two broad categories: natural and synthetic.

Natural fertilizers are commonly misnamed "organic." "Natural fertilizers" is a more accurate description because these materials can be both complex chemical substances containing carbon (organic materials) or inorganic ores, such as rock phosphate, which are mined. Natural fertilizers containing organic materials include manures and composts, animal byproducts (such as bone meal, blood meal, feather meal), and seed meals. Natural fertilizers that are inorganic ores include potassium and lime.

Natural fertilizers typically release nutrients at a slower rate and over a longer period than synthetic fertilizers because microorganisms are involved in a breakdown and release cycle called mineralization. Moisture, temperature, and the microbial species and populations in the soil affect mineralization. Some water-soluble natural fertilizers, such as fish emulsion, are available when rapid nutrient delivery is desired. When using natural fertilizers, it is helpful to incorporate them and provide adequate moisture for active microbial populations.

When packaged as fertilizers, natural fertilizers will have the nutrient analysis stated on the labels. How much to use varies with the nutrient content of the material. The age of the material is also a factor. Producers are not required by law to state the nutrient content on bulk organic materials, such as compost, manure, and sludges. The source of these materials should be investigated and possible analysis performed at the Plant, Waste, Solution, and Media Lab at the NCDA&CS Agronomic Division before applying large amounts to a home garden. The age of the natural fertilizers is another important factor. When natural material decays and is rained on, it loses nutrients, especially potassium and, to some extent, nitrogen. Even natural sources of nutrients can be overapplied and damage plants. Fresh manures, for example, may injure plants by adding excessive nitrogen or potassium, especially when applied in large quantities.

Natural fertilizers can be expensive if applied in amounts adequate to supply nutrients for good plant growth, but have the added benefit of improving soil structure and plant vigor. When applying natural fertilizers, calculate as closely as possible the amounts of nutrients being supplied. Always err on the low side of application rates, then test the soil and augment as recommended on the soil test report. The nutrient content may need to be supplemented with other natural or synthetic materials to achieve a balanced ratio of nutrients.

Synthetic fertilizers are made through industrial processes or mined from deposits in the earth. They are purified, mixed, blended, and altered for easy handling and application. Most are non-carbonaceous chemicals from non-living sources and are usually cheaper than natural fertilizers. In general, nutrients are more rapidly available to plants because they are more water-soluble or in a form plants can use. The disadvantage is that it may be easier to over-apply a synthetic fertilizer than a natural one, which may result in fertilizer burn. In addition, synthetic fertilizers may not support beneficial microbial populations to the same extent as natural fertilizers.

Special-purpose fertilizers are packaged for plants such as camellias, rhododendrons, and azaleas. Some of the compounds used in these fertilizers have an acid reaction that can be beneficial to acid-loving plants if the soil they are growing in is naturally neutral or alkaline; however, most soils in North Carolina are usually acidic so these special fertilizers are unnecessary. Fertilizer spikes or pellets are fertilizers compressed into a form placed in the soil or pots. They are convenient, but are expensive per unit of fertilizer and do not provide uniform distribution. Nutrients are often concentrated around the spikes or pellets. Liquid fertilizer can be purchased as a dry powder or as a concentrated liquid.

Liquid fertilizers are frequently used for houseplants or as a starter solution for transplants. They tend to be more expensive per unit of fertilizer because they are made from refined chemicals. Foliar fertilizers are dry powders or concentrated liquids that are mixed with water and sprayed on plants. Foliar feeding is used when insufficient fertilizer was applied before planting, when a quick growth response is wanted, when micronutrients are locked in the soil, or when the soil is too cold for the plant to use fertilizer in the soil. Foliar-applied nutrients are absorbed and used by the plant quite rapidly. They are expensive per unit of nutrient and only give short-term fertilization (completely absorbed within one to two days). Relying totally on foliar fertilization can be time consuming because the fertilizer must be applied regularly. Improper foliar application of fertilizers can also lead to plant tissue burn.

Incomplete fertilizers can be used separately or combined to supply the needed nutrients, often at a reduced cost compared to using a complete fertilizer. For example, gardeners who have a soil with sufficient P and K can save money by applying a nitrogen-only fertilizer, such as ammonium nitrate (34-0-0). If a soil test indicates N and K are needed, but not P, use an appropriate amount of ammonium nitrate and muriate of potash (0-0-60), a naturally occurring material composed almost entirely of potassium, processed to remove impurities and concentrate the product. If a soil needs only P, use triple super phosphate (0-46-0), or for an organic nutrient source apply bone meal (approximately 3-15-0; note that this will add some N) or compost.

Plant Nutrients and the Environment

Fertilizer misuse causes environmental and water quality issues. Nitrogen fertilizers, for instance, break down into ammonium and nitrate. The nitrate form of N, while essential for plant growth, is highly mobile and can move through the soil after rainfall or irrigation and contaminate drinking water supplies. Phosphorus holds tightly to soil particles and does not leach through the soil, but affects water quality through runoff and soil erosion. Excess nitrogen and phosphorus are associated with algal blooms (heavy growth of aquatic plants) and limited oxygen, and cause fish kills in lakes, bays, and non-flowing water bodies.

There are several ways to reduce fertilizers' impacts on water quality:

- Apply only materials that are recommended based on results of a soil test. If possible, use slow-release fertilizers and incorporate into the soil. Avoid applying excess nitrogen and phosphorus fertilizer.
- Calibrate fertilizer spreaders properly and clean spreaders over the lawn area instead of a hard surface.
- Keep the amount of hard surfaces in a landscape to a minimum. When installing a new sidewalk or patio, consider using gravel, porous concrete, stepping-stones, wood decking, or bricks on a sand base.
- Avoid applying fertilizer to hard surfaces, such as sidewalks, patios, driveways, and streets. Sweep up material that falls on hard surfaces.
- Maximize water absorption by aerating lawns and incorporating organic matter in planting beds and gardens.
- Prevent runoff by turning off irrigation when the soil is no longer absorbing water.
- Avoid applying fertilizer in natural drainage areas or ditches.
- Minimize soil erosion by using ground covers, windbreaks, terraces, and mulches.
- Mulch under trees and shrubs to reduce impact of falling water.
- Maintain a lawn border around planting areas and plant a grass strip between rows in fruit and vegetable gardens.
- Plant cover crops on bare soil, such as barren vegetable gardens.
- Use a rain barrel under drains to collect runoff and direct excess runoff from roofs onto grassy areas.

When to Apply Fertilizer

Soil type affects the frequency of fertilizer application. Sandy soils require more frequent applications of smaller amounts of nitrogen and potash than do clay soils because these nutrients leach more readily in sandy soils. Other factors that affect application frequency include the plant to be grown, the amount of plant growth desired, the amount of water, and the type and release rate of fertilizer applied.

The best time to apply fertilizer and the most effective method of applying it depends on the type of plants being grown. Leafy vegetables require more nitrogen than root crops. Corn is a heavy nitrogen feeder and may require several small nitrogen applications when actively growing. Most established woody plants perform well without fertilization, or with just one application per year. Young plants may benefit from several light applications of fertilizer per year.

Fertilizer is needed when plants are actively growing, never when they are dormant. Nitrogen application will have its greatest effect three to four weeks after application. Excess or improperly timed nitrogen can delay flowering and fruiting or promote tender new growth vulnerable to frost or freeze damage. Research has shown it is best to broadcast or incorporate fertilizer uniformly over an area rather than concentrating fertilizer in holes or bands in the soil. The most effective method of fertilizing a large area is with a fertilizer spreader; for home gardens, hand fertilization works fine. For new plantings, incorporate fertilizer into the soil and mix it thoroughly. For established plantings, surface application is appropriate. When fertilizing from overhead, make certain plant foliage is dry and use a broom to brush fertilizer off the foliage, or water thoroughly after applying fertilizer to remove it from plant leaves to prevent burn spots. It is not necessary to remove mulch when fertilizing; irrigation or rainfall will carry fertilizer to the roots. Fertilization should be reduced or delayed during dry weather because the salts in the fertilizer can burn roots if there is inadequate moisture.

Fertilizer Terms

Fertilizer analysis: The minimum amount of each element in a fertilizer as stated on the label, such as 16-4-8.

Fertilizer ratio: The relative proportion of N, P2O5, and K2O. The ratios of 16-4-8 and 8-2-4 are both 4:1:2, which means 4 parts nitrogen to 1 part phosphorus to 2 parts potassium.

Balanced fertilizer: A fertilizer containing equal parts of each major element, such as 10-10-10.

Complete fertilizer: A fertilizer containing nitrogen, phosphorus, and potassium. Examples of commonly used fertilizers are 10-10-10, 16-4-8, and 12-4-8.

Incomplete fertilizer: A fertilizer missing one or two of the macronutrients, such as 0-20-0.

Weed and feed fertilizers: A combination of fertilizer and herbicide. They are often used on lawns to prevent certain weeds from germinating, or to kill existing broadleaf weeds.

High analysis: A fertilizer containing 30% or more active nutrients, such as ammonium nitrate 33-0-0. The cost per bag is usually more, but the cost per pound of nutrient is less, lowering the cost for fertilizing a given area.

Figure: *A fertilizer specially formulated for acidic plants such as azaleas or rhododendrons. These fertilizers are usually unnecessary in North Carolina's acidic soils.*

Figure: *Fertilizer spikes are compressed fertilizer that can be pushed into the soil.*

Figure: *Liquid fertilizer can be found at any local garden center.*

Figure: *Foliar fertilizers are dry powders mixed with liquid and then sprayed on the leaves of plants.*

Figure: *This homeowner used permeable pavers for their driveway allowing grass to grow in the spaces and reducing runoff from the property.*

Figure: *A rain barrel catches runoff from a roof top and can be used to water plants in the landscape.*

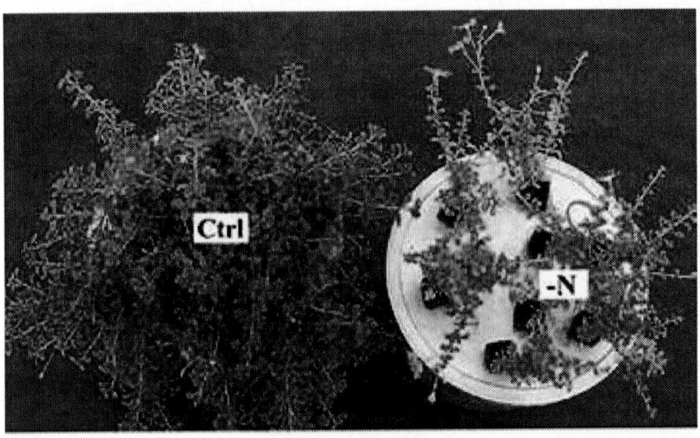

Figure: *Nitrogen deficiency in Bacopa. Note the severely reduced growth and light green-yellow leaves.*

Figure: *Potassium deficiency in Dracaena halapepe.*

Figure: *Magnesium deficiency appears first in older leaves. On this bean plant, the veins remain green while the rest of the leaf turns yellow.*

Figure: *Calcium deficiency in tomatoes.*

Figure: *Sulfur deficiency rarely occurs, but on this strawflower, it has affected the shoot size and leaves with a yellowing of the veins.*

***Figure:** Iron deficiency causing interveinal chrolorosis on young leaves.*

***Figure:** This geranium is deficient in boron which is evident by the thick, leathery, chlorotic younger leaves.*

***Figure:** Boron toxicity on a rose bush. Can look like the inverted "V" of green on a yellow leaf that is characteristic of magnesium deficiency, but uptake of too much boron by the plant often occurs when pH is too low.*

Figure: Copper deficiency in this daisy appears as small misshapen growth.

***Figure:** Manganese deficiency in a chrysanthemum exhibited as reduced growth while the youngest leaves remain green.*

***Figure:** Molybdenum deficiency on poinsettia.*

Soil pH and Nutrient Availability

The desirable soil pH range for optimum plant growth varies among crops. Generally, soil pH 6.0-7.5 is acceptable for most plants as most nutrients become available in this pH range. Soil pH can be determined by mixing soil sample with water and then measuring the resulting aqueous solution.

Soil pH is a measure of the acidity or alkalinity in soils. In the pH scale, pH 7.0 is neutral. Below 7.0 is acidic and above 7.0 is basic or alkaline. Soil pH affects nutrients available for plant growth. In highly acidic soil, aluminum and manganese can become more available and more toxic to plant while calcium, phosphorus, and magnesium are less available to the plant. In highly alkaline soil, phosphorus and most micronutrients become less available.

When designing or planting new garden or landscape, it is helpful to check the soil pH as different plants thrive in different soil pH ranges. pH determination can give indication whether soil is suitable for the plants to be grown or it needs to be adjusted to produce optimum plant growth.

Soil pH can be measured easily and inexpensively at home or on-site using LAQUAtwin pH meters. There are three (3) LAQUAtwin pH meter models available, namely pH 11, 22, and 33. These pocket-sized compact meters allow two to five calibration points using either NIST or USA pH buffers. Among the above-mentioned models, the pH 33 meter has built-in temperature sensor that measures and displays temperature reading and automatic temperature compensation feature (ATC) that performs automatic calibration to the exact pH of the buffer at the measured temperature. Refer to the specifications of each meter model for more information.

Method

Calibrate the LAQUAtwin pH meter according to manufacturer's instructions using at least two pH buffers that bracket the expected sample pH.

Sample Preparation and Measurement

The method described below is based on US EPA Method 9045D. This is also applicable for measuring pH of waste samples, which may be solids, sludges, or non-aqueous liquids. If water is present, it must constitute less than 20% of the total volume of the sample.

Add 20ml of pure water to 20g sample in a beaker or container. Stir for 5 minutes then cover.

Let the soil suspension stand for about 1 hour. Alternatively, filter or centrifuge off the water phase.

Measure the pH of water phase. Record the pH value and the temperature.

To obtain accurate results, standard buffer solutions and samples should be measured at the same temperature. If the electrode is coated with oily material from a sample, clean it with detergent and warm water.

Results and Benefits

The desirable soil pH range for optimum plant growth varies among crops. Generally, soil pH 6.0-7.5 is acceptable for most plants as most nutrients become available in this pH range. Soil pH is important because it affects the availability of nutrients to plants. Nitrogen, phosphorus, and potassium are the primary nutrients needed in fairly large quantities.

Calcium, magnesium, and sulfur are secondary nutrients required by the plant in lesser quantities. Zinc and manganese are micronutrients required by the plant in very small amounts. Most secondary and micronutrient deficiencies are easily corrected by keeping the soil at the optimum pH value.

Soil pH also affects activity of soil microorganisms. The population of bacteria that decompose organic matter declines and their activity is hindered in highly acidic soil, which results in accumulation of organic matter and the bound nutrients, particularly nitrogen.

Increasing Soil Ph

Applying a material that contains some form of lime (calcium carbonate) like the ground agricultural limestone and wood ashes can increase the soil pH. The finer the limestone, the more rapid it becomes effective. Different soils will require a different amount of lime to adjust the soil pH. Wood ashes contain high amounts of potassium and calcium, and small amounts of phosphate, boron and other nutrients. Although not as effective as limestone, these can drastically increase the soil pH with repeated use.

Decreasing Soil pH

Aside from ammonium-based fertilizers and organic matter, aluminum sulfate and sulfur are common materials used for decreasing soil pH. Aluminum sulfate is preferred as it changes the soil pH as soon as it dissolves in the soil because of the aluminum. However, too much of this is toxic to plant. Sulfur takes some time to produce effect as it needs to be converted to sulfuric acid by soil bacteria

Soil Organic Matter: Role, Dynamics and Management

Soil Organic Matter

When plant residues are returned to the soil, various organic compounds undergo decomposition. Decomposition is a biological process that includes the physical breakdown and biochemical transformation of complex organic molecules of dead material into simpler organic and inorganic molecules (Juma, 1998).

The continual addition of decaying plant residues to the soil surface contributes to the biological activity and the carbon cycling process in the soil. Breakdown of soil organic matter and root growth and decay also contribute to these processes. Carbon cycling is the continuous transformation of organic and inorganic carbon compounds by plants and micro- and macro-organisms between the soil, plants and the atmosphere.

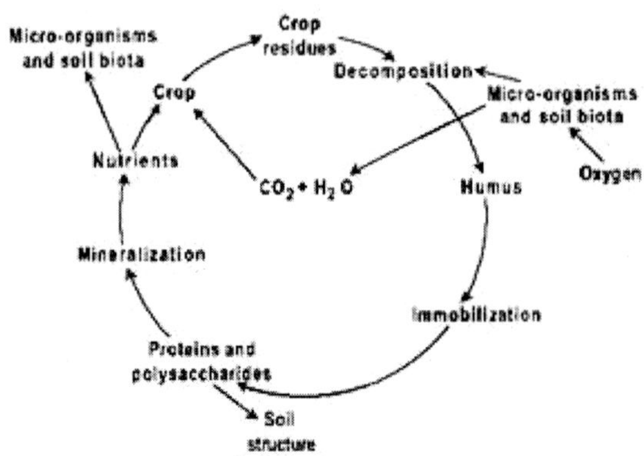

Figure: *Carbon cycle*

Decomposition of organic matter is largely a biological process that occurs naturally. Its speed is determined by three major factors: soil organisms, the physical environment and the quality of the organic matter. In the decomposition process, different products are released: carbon dioxide (CO2), energy, water, plant nutrients and resynthesized organic carbon compounds. Successive decomposition of dead material and modified organic matter results in the formation of a more complex organic matter called humus. This process is called humification. Humus affects soil properties. As it slowly decomposes, it colours the soil darker; increases soil aggregation and aggregate stability; increases the CEC (the ability to attract and retain nutrients); and contributes N, P and other nutrients. Soil organisms, including micro-organisms, use soil organic matter as food. As they break down the organic matter, any excess nutrients (N, P and S) are released into the soil in forms that plants can use.

This release process is called mineralization. The waste products produced by micro-organisms are also soil organic matter. This waste material is less decomposable than the original plant and animal material, but it can be used by a large number of organisms. By breaking down carbon structures and rebuilding new ones or storing the C into their own biomass, soil biota plays the most important role in nutrient cycling processes and, thus, in the ability of a soil to provide the crop with sufficient nutrients to harvest a healthy product. The organic matter content, especially the more stable humus, increases the capacity to store water and store (sequester) C from the atmosphere.

The Soil Food Web

The soil ecosystem (Box 1) can be defined as an interdependent life-support system composed of air, water, minerals, organic matter, and macro- and micro-organisms, all of which function together and interact closely. The organisms and their interactions enhance many soil ecosystem functions and make up the soil food web. The energy needed for all food webs is generated by primary producers: the plants, lichens, moss, photosynthetic bacteria and algae that use sunlight to transform CO2 from the atmosphere into carbohydrates. Most other organisms depend on the primary producers for their energy and nutrients; they are called consumers.

Soil life plays a major role in many natural processes that determine nutrient and water availability for agricultural productivity. The primary activities of all living organisms are growing and reproducing. By-products from growing roots and plant residues feed soil organisms. In turn, soil organisms support plant health as they decompose organic matter, cycle nutrients, enhance soil structure and control the populations of soil organisms, both beneficial and harmful (pests and pathogens) in terms of crop productivity.

The living part of soil organic matter includes a wide variety of micro-organisms such as bacteria, viruses, fungi, protozoa and algae. It also includes plant roots, insects, earthworms, and larger animals such as moles, mice and rabbits that spend part of their life in the soil. The living portion represents about 5 percent of the total soil organic matter.

Micro-organisms, earthworms and insects help break down crop residues and manures by ingesting them and mixing them with the minerals in the soil, and in the process recycling energy and plant nutrients. Sticky substances on the skin of earthworms and those produced by fungi and bacteria help bind particles together. Earthworm casts are also more strongly aggregated (bound together) than the surrounding soil as a result of the mixing of organic matter and soil mineral material, as well as the intestinal mucus of the worm. Thus, the living part of the soil is responsible for keeping air and water available, providing plant nutrients, breaking down pollutants and maintaining the soil structure.

The composition of soil organisms depends on the food source (which in turn is season dependent). Therefore, the organisms are neither uniformly distributed through the soil nor uniformly present all year. However, in some cases their biogenic structures remain. Each species and group exists where it can find appropriate food supply, space, nutrients and moisture (Plate 2). Organisms occur wherever organic matter occurs (Ingham, 2000). Therefore, soil organisms are concentrated: around roots, in litter, on humus, on the surface of soil aggregates and in spaces between aggregates. For this reason, they are most prevalent in forested areas and cropping systems that leave a lot of biomass on the surface.

Figure: *Termites create their own living conditions near their preferred food sources. Inside the colony life is highly organized*

The activity of soil organisms follows seasonal as well as daily patterns. Not all organisms are active at the same time. Most are barely active or even dormant.

Availability of food is an important factor that influences the level of activity of soil organisms and thus is related to land use and management.

Practices that increase numbers and activity of soil organisms include: no tillage or minimal tillage; and the maintenance of plant and annual residues that reduce disturbance of soil organisms and their habitat and provide a food supply.

Different groups of organisms can be distinguished in the soil.

Decomposition Process

Fresh residues consist of recently deceased micro-organisms, insects and earthworms, old plant roots, crop residues, and recently added manures.Crop residues contain mainly complex carbon compounds originating from cell walls (cellulose, hemicellulose, etc.). Chains of carbon, with each carbon atom linked to other carbons, form the "backbone" of organic molecules.

These carbon chains, with varying amounts of attached oxygen, H, N, P and S, are the basis for both simple sugars and amino acids and more complicated molecules of long carbon chains or rings. Depending on their chemical structure, decomposition is rapid (sugars, starches and proteins), slow (cellulose, fats, waxes and resins) or very slow (lignin).

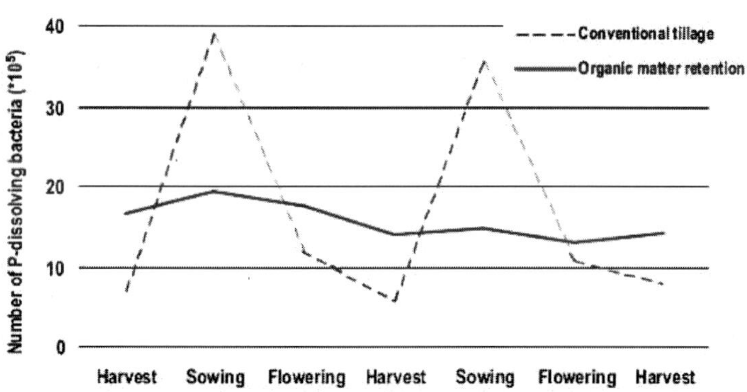

***Figure:** Fluctuations in microbial biomass at different stages of crop development in conventional agriculture compared with systems with residue retention and high organic matter input*

Decrease in Biomass Production

Replacement of Perennial Vegetation

A consequence of clearing forest for agriculture is the disappearance of the litter layer, with a consequent reduction in the numbers and variety of soil organisms. While many temperate forest species appear to adapt well to grassland, the effects of deforestation in the tropics appear to be more marked. Studies have shown that as soil biodiversity declines, adapted species may take over from the indigenous species and the composition may change drastically.Soil macrofaunal biomass and population density fell to 6 and 17 percent, respectively, in cultivated plots, compared with primary forest in Peruvian Amazonia. In Suriname, the number of animals per square metre has fallen to 36 percent and the diversity of species has fallen to 28 percent compared with primary forest. The indigenous species have largely disappeared, but adapted species have been available for recolonization. The composition of the macrofaunal community has changed drastically.

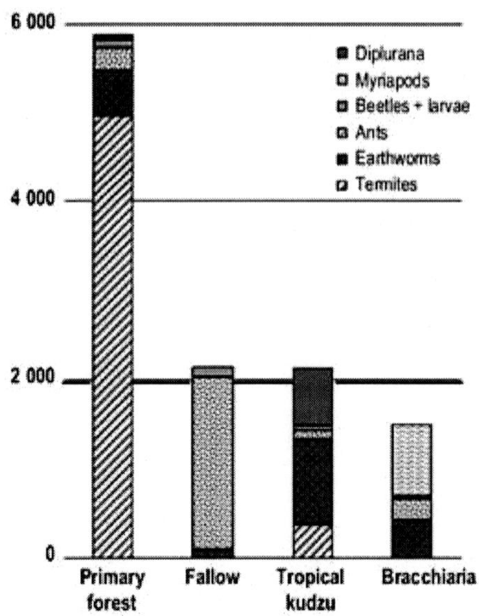

Figure: *Composition of soil macrofauna under primary forest, fallow, kudzu and grass vegetation*

Replacement of mixed vegetation with monoculture of crops and pasturesThe simplification of vegetation and the disappearance of the litter layer under grassland and monocrop production systems lead to a decrease in faunal diversity. Although root systems (especially of grasses) can be extensive and explore vast areas of soil, the root exudates from one single crop will attract only a few different microbial species. This in turn will affect the predator diversity. The more opportunistic pathogen species will be able to acquire space near the crop and cause harm. Continuous cultivation and grazing also leads to compaction of soil layers, which in turn affects the circulation of air. Anaerobic conditions in the soil stimulate the growth of different micro-organisms, resulting in more pathogenic organisms.

High harvest index: One of the consequences of the green revolution was the replacement of indigenous varieties of species with high-yielding varieties (HYVs). These HYVs often produce more grain and less straw, compared with locally developed varieties; the harvest index of the crop (ratio of grain to total plant mass aboveground) is increased. From a production point of view, this is a logical approach. However, this is less desirable from a conservation point of view. Reduced amounts of crop residues remain after harvest for soil cover and organic matter, or for grazing of livestock (which results in manure). Moreover, where animals graze the residues, even less remains for conservation purposes.

Use of bare fallow: Traditionally, a fallow period is used after a period of crop production to give the land some "rest" and to regenerate its original state of productivity. Usually, this is necessary in production systems that have drawn down the nutrient supply and altered the soil biota significantly, such as in slash-and-burn systems or conventional tillage systems.

Some farmers use bare fallow to regenerate their lands. However, apart from spontaneous weed growth, this means there is no energy source for the soil biota present on the land. Instead of recovering the soil food web, the soil organic matter is degraded further and the lack of cover can result in severe erosion and runoff when the rains start after the dry season.

Decrease In Organic Matter Supply

Burning of natural vegetation and crop residues: The burning of maize, rice and other crop residues in the field is a common practice (Plate 5). Residues are usually burned to help control insects or diseases or to make fieldwork easier in the following season. Burning destroys the litter layer and so diminishes the amount of organic matter returned to the soil. The organisms that inhabit the surface soil and litter layer are also eliminated. For future decomposition to take place, energy has to be invested first in rebuilding the microbial community before plant nutrients can be released. Similarly, fallow lands and bush are burned before cultivation. This provides a rapid supply of P to stimulate seed germination. However, the associated loss of nutrients, organic matter and soil biological activity has severe long-term consequences.

Overgrazing: There is a tendency throughout the world to overstock grazing land above its carrying capacity. Cows, draught animals and small ruminants graze on communal grazing areas and on roadsides, stream banks and other public land (Plate 6). Overgrazing destroys the most palatable and useful species in the plant mixture and reduces the density of the plant cover, thereby increasing the erosion hazard and reducing the nutritive value and the carrying capacity of the land.

Removal of crop residues: Many farmers remove residues from the field for use as animal feed and bedding or to make compost (Plate 7). Later, these residues return to contribute to soil fertility as manures or composts. However, residues are sometimes removed from the field and not returned. This removal of plant material impoverishes the soil as it is no longer possible to recycle the plant nutrients present in the residues.

Figure: *Animal at pasture in the area around Asmara.*

Increased Decomposition Rates

Tillage Practices

Tillage is one of the major practices that reduces the organic matter level in the soil. Each time the soil is tilled, it is aerated. As the decomposition of organic matter and the liberation of C are aerobic processes, the oxygen stimulates or speeds up the action of soil microbes, which feed on organic matter.

This means that:

- When ploughed, the residues are incorporated in the soil together with air and come into contact with many micro-organisms, which accelerates the carbon cycle. The decomposition is faster, resulting in the formation of less stable humus and an increased liberation of CO2 to the atmosphere, and thus a reduction in organic matter.
- The residues on the soil surface slow the carbon cycle because they are exposed to fewer micro-organisms and thus wane more slowly, resulting in the production of humus (which is more stable), and liberating less CO2 to the atmosphere.

Figure: *Removed crop residues cannot serve as food for soil organisms.*

TABLE Tillage induced flush of decomposition of organic matter

Type of tillage	Organic matter lost in 19 days (kg/ha)
Mouldboard plough + disc harrow (2x)	4 300
Mouldboard plough	2 230
Disc harrow	1 840
Chisel plough	1 720
Direct seeding	860

In terms of short-term organic matter loss, the more a soil is tilled, the more the organic matter is broken down. There are also longer-term losses, attributed to repeated, annual cultivation. Cropping systems that return little residue to the soil accelerate this decline. Many modern cropping systems combine frequent tillage with small amounts of residue, with resultant reductions in the organic matter content of many soils.

Historically, manure application (from farm livestock) was common, and it was a dynamic way of maintaining organic matter levels despite repeated cultivation and low residue returns to the soil. Increased on-farm mechanization has reduced livestock numbers, so this source of organic material has been reduced considerably.

Organic matter production and conservation is affected dramatically by conventional tillage, which not only decreases soil organic matter but also increases the potential for erosion by wind and water.

The impact occurs in many ways:

- Ploughing leaves no residues on the soil surface to lessen the impact of rain.
- Ploughing reduces the quantity of food sources for earthworms and disturbs their burrows and living space, hence populations of certain species decrease drastically. Moreover, reduction of earthworm numbers reduces their impact, through burrowing, in increasing porosity and aeration (particularly continuous macropores) and lowers their ability to bury and incorporate plant residues, which facilitates rapid decomposition of organic matter.
- Tillage by repeated hoeing or discing smoothes the surface and destroys natural soil aggregates and channels that connect the surface with the subsoil, leaving the soil susceptible to erosion. Old root channels and earthworm holes are eliminated, as are the cracks between natural aggregates. The large pores, the ones destroyed by conventional tillage practices, are necessary to conduct water into the soil during rainfall.
- The development of a plough pan or hoe pan, a layer of compacted soil resulting from smearing action at the bottom of the plough or hoe, may retard both root penetration and water infiltration.
- Ploughing or discing under dry conditions exacerbates the pulverization of the soil, causing the soil surface to crust more easily, leading to greater water runoff and erosion. This is exacerbated by reduced soil surface roughness, which leaves few depressions for temporary storage of water during intense storms.
- Increased runoff during rainstorms may also increase the possibility of drought stress later in the season, because water that runs off the field does not infiltrate into the soil to remain available to plants.

In some circumstances, imbalances of certain soil organisms can disrupt soil structure and processes, e.g. certain earthworm species in rice fields or pastures.

Figure: *Intensive soil tillage makes the land vulnerable to erosive processes, as the organic matter is lost through increased oxidation in the soil, the upper subsoil is ompact, and the loosened topsoil can more readily wash away.*

Drainage: Decomposition of organic matter occurs more slowly in poorly aerated soils, where oxygen is limiting or absent, compared with well-aerated soils. For this reason, organic matter accumulates in wet soil environments. Soil drainage is determined strongly by topography - soils in depressions at the bottom of hills tend to remain wet for extended periods of time because they receive water (and sediments) from upslope. Soils may also have a layer in the subsoil that inhibits drainage, again exacerbating waterlogging and reduction in organic matter decomposition. In a permanently waterlogged soil, one of the major structural parts of plants, lignin, does not decompose at all. The ultimate consequence of extremely wet or swampy conditions is the development of organic (peat or muck) soils, with organic matter contents of more than 30 percent. Where soils are drained artificially for agricultural or other uses, the soil organic matter decomposes rapidly.

Fertilizer and pesticide use: Initially, the use of fertilizer and pesticides enhances crop development and thus production of biomass (especially important on depleted soils). However, the use of some fertilizers, especially N fertilizers, and pesticides can boost micro-organism activity and thus decomposition of organic matter. The chemicals provide the microorganisms with easy-to-use N components. This is especially important where the C: N ratio of the soil organic matter is high and thus decomposition is slowed by a lack of N.

Practices That Increase Soil Organic Matter: Increased concern about the environmental and economic impacts of conventional crop production has stimulated interest in alternative systems. Central to such systems is the need to promote and maintain soil biological processes and minimize fossil fuel inputs in the form of fertilizers, pesticides and mechanical cultivation. All activities aimed at the increase of organic matter in the soil (Box 3) help in creating a new equilibrium in the agro-ecosystem. For a system of natural resource management to be balanced, and thus sustainable, it must be able to withstand sharp climatic fluctuations, and to evolve steadily in response to social changes and changes in the costs and availability of inputs of land, labour and knowledge.

The more diverse and complex an agricultural system is, the more stable and sustainable it will be in the face of unpredictable vagaries of climate and market. Thus, annual crops, woody perennials and nonwoody perennials may be combined in various ways with livestock or trees, or both, in what are now commonly called agrosilvipastoral systems. Different approaches are required for different soil and climate conditions.

However, the activities will be based on the same principle: increasing biomass production in order to build active organic matter. Active organic matter provides habitat and food for beneficial soil organisms that help build soil structure and porosity, provide nutrients to plants, and improve the water holding capacity of the soil. Several cases have demonstrated that it is possible to restore organic matter levels in the soil (Figure 8). Activities that promote the accumulation and supply of organic matter, such as the use of cover crops and refraining from burning, and those that reduce decomposition rates, such as reduced and zero tillage, lead to an increase in the organic matter content in the soil (Sampson and Scholes, 2000).

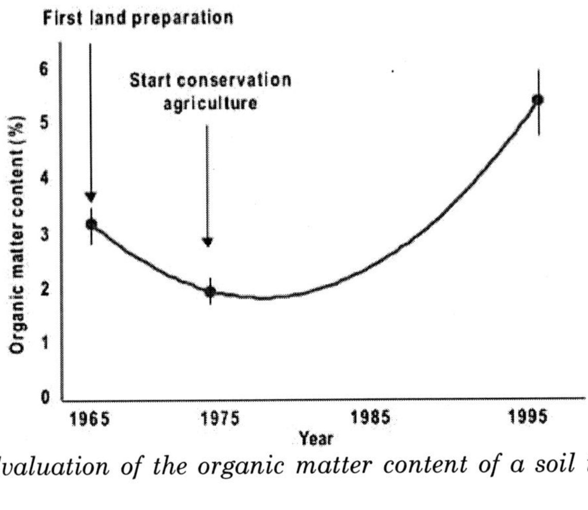

Figure: *Evaluation of the organic matter content of a soil in Paraná*

Increased Biomass Production

Increased water availability for plants: water harvesting and irrigation

In dry conditions, water may be provided through irrigation or water harvesting. The increased water availability enhances biomass production, soil biological activity and plant residues and roots that provide organic matter.

The concept of water harvesting includes various technologies for runoff management and utilization. It involves capture of runoff (in some cases through treating the upstream capture area), and its concentration on a runon area for use by a specific crop (annual or perennial) in order to enhance crop growth and yields, or its collection and storage for supplementary irrigation or domestic or livestock purposes. The objective of designing a water harvesting system is to obtain the best ratio of the area yielding runoff to either the area where runoff is being directed or the capacity of the storage structure (volume of water collected).

Figure: *Hollows are dug in the ground to gather water during the rainy season.*

In this way, the water captured for crop production during runoff periods can be stored either directly in the soil for subsequent use by plants or in small farm reservoirs or collection tanks. This aids stabilization of crop production by enhancing soil moisture availability or allowing irrigation during a dry period within the rainy season or by extending crop production into the dry season. Some factors to be considered regarding these runoff farming systems and reservoirs include: site selection, watershed size and condition, rainfall distribution and runoff, and water requirements of crops. Where a minimum water depth of about 1 m can be maintained in a reservoir, fish can be raised to provide additional food. Numerous water harvesting systems have been developed over the centuries, especially in arid areas. The principle of collecting runoff for crop production is also inherent to many other soil and water conservation technologies that apply the concept of runoff and runon areas at a microwatershed level, such as negarims, trapezoidal or "eyebrow" bunds and tied ridges.

Balanced fertilization: Where the supply of nutrients in the soil is ample, crops are more likely to grow well and produce large amounts of biomass. Fertilizers are needed in those cases where nutrients in the soil are lacking and cannot produce healthy crops (FAO, 2000) and sufficient biomass. Most soils in sub-Saharan Africa (SSA) are deficient in P. P is required not only for plant growth but also for N fixation. Unbalanced fertilization, for example mainly with N, may result in more weed competition, higher pest incidence and loss of quality of the product. Unbalanced fertilization eventually leads to unhealthy plants. Therefore, fertilizers should be applied in sufficient quantities and in balanced proportions. The efficiency of fertilizer use will be high where the organic matter content of the soil is also high. In very poor or depleted soils, crops use fertilizer applications inefficiently. When soil organic matter levels are restored, fertilizer can help maintain the revolving fund of nutrients in the soil by increasing crop yields and, consequently, the amount of residues returned to the soil.

Cover crops: Growing cover crops is one of the best practices for improving organic matter levels and, hence, soil quality. The benefits of growing cover crops include:

- They prevent erosion by anchoring soil and lessening the impact of raindrops.
- They add plant material to the soil for organic matter replenishment.
- Some, e.g. rye, bind excess nutrients in the soil and prevent leaching.
- Some, especially leguminous species, e.g. hairy vetch, fix N in the soil for future use.
- Most provide habitat for beneficial insects and other organisms.
- They moderate soil temperatures and, hence, protect soil organisms.

A range of crops can be used as vegetative cover, e.g. grains, legumes and oil crops. All have the potential to provide great benefit to the soil. However, some crops emphasize certain benefits; a useful consideration when planning a rotation scheme. It is important to start the first years with (cover) crops that cover the surface with a large amount of residues that decompose slowly (because of the high C:N ratio). Grasses and cereals are most appropriate for this stage, also because of their intensive rooting system, which improves the soil structure rapidly. In the following years, when soil health has begun to improve, legumes can be incorporated in the rotation. Leguminous crops enrich the soil with N and their residues decompose rapidly because of their low C:N ratio. Later, when the system is stabilized, it is possible to include cover crops with an economic function, e.g. livestock fodder. The selection of cover crops should depend on the presence of high levels of lignin and phenolic acids. These give the residues a higher resistance to decomposition and thus result in soil protection for a longer period and the production of more stable

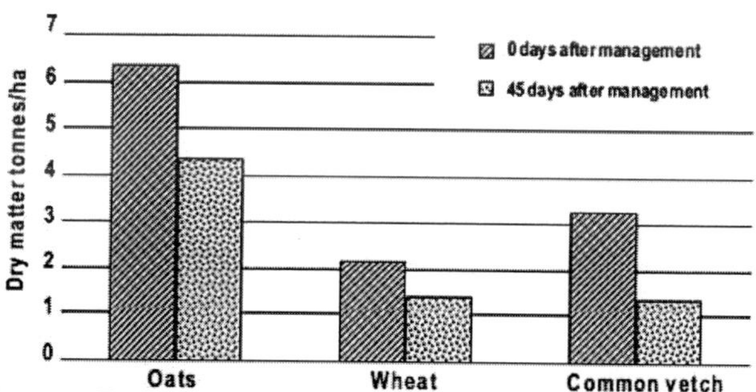

Figure: *Reduction of dry matter of different cover crops*

Source: Ruedell, 1995 humus.

Another determining factor in the dynamics of residue composition is the biochemical composition of the residues. Depending on species, their chemical components and the time and way of managing them, there will be differences in decomposition rates. The grain species (oats and wheat) show more resistance than common vetch (legume) to decomposition. The latter has a lower C:N ratio and a lower lignin content and is thus subject to a rapid decomposition.

Agricultural production systems in which residues are left on the soil surface, such as direct seeding and the use of cover crops, stimulate the development and activity of soil fauna at many levels.

The term green manure is often used to indicate the same plant species that are used as cover crops. However, green manure refers specifically to a crop in the rotation grown for incorporation of the non-decomposed vegetative matter in the soil. While this practice is used specifically to add organic matter, this is not the most effective use of organic matter (especially in hot climates) for two reasons:

- Mechanical disturbance of the soil should be avoided as much as possible.
- When biomass is incorporated in the soil all at one time, there is a short period of high microbial activity in decomposing the material. This results in the sudden release of a large quantity of nutrients that cannot be captured by the seedlings of the following crop and is thus lost from the system.

In general, the greater the production of green manure or crop biomass, the greater is the microbial, mesofauna and macrofauna population of the soil - from fungi and micro-organisms to earthworms and termites. The dynamics of surface residue decomposition depend inter alia on the activity of micro-organisms and also on soil mesofauna and macrofauna. The macrofauna consists mainly of earthworms, beetles, termites, ants, millipedes, spiders, snails and slugs. These organisms help integrate the residues into the soil and improve soil structure, porosity, water infiltration, and through-flow through the creation of burrows, ingestion and secretions.

The natural incorporation of cover-crop and weed residues from the soil surface to deeper layers in the soil by soil macrofauna is a slow process. The activity of microorganisms is regulated by the activity of the macrofauna, because the latter provide them with food and air through their bioturbation activities. In this way, nutrients are released slowly and can provide the crop with nutrients over a longer period. At the same time, the soil is covered for a long time by the residues and is protected against the impact of rain and sun.

Improved vegetative stands: In many places, low plant densities limit crop yields. Wide plant spacing is often practised as "a way to return power to the soil" or "to give the soil some rest", but in reality it is an indicator that the soil is impoverished. Plant spacing is usually determined by farmers in relation to soil fertility and available water or expected rainfall (unless standard recommendations are enforced by extension). This means that plants are often spaced widely on depleted soils in arid and semi-arid regions with a view to ensuring an adequate provision of plant nutrients and water for all plants.

However, it is important to maintain the recommended plant spacing in order to optimize biomass production and rooting density and, hence, organic matter for food, moisture retention and habitat for soil organisms. Once the crop is established, reduced sunlight between closer crop rows may also reduce regrowth of weeds.

Planting pits are a way of increasing biomass production and crop yields on severely degraded land in semi-arid conditions. Rainfall is concentrated near the plants, and soil faunal activity and organic matter accumulation are concentrated in the planting pits. Planting pits have been introduced successfully in Zambia as a conservation practice for smallholder farmers, who do not have fertilizers or tractor services available to them.

Figure: *Half-moons around newly planted Acacia seedlings catch and retain rainwater. Agroforestry and alley cropping*

Agroforestry is a collective name for land-use systems where woody perennials (trees, shrubs, palms, etc.) are integrated in the farming system. Alley cropping is an agroforestry system in which crops are grown between rows of planted woody shrubs or trees. These are pruned during the cropping season to provide green manure and to minimize shading of crops.

Agroforestry covers a wide range of systems combining food crops, forestry and pasture species in different ways (agrosilviculture, silvipasture, agrosilvipasture and multipurpose forest production). There are two different approaches to agroforestry. One uses agricultural crops or pasture as a transitional means of utilizing the land until forest plantations are fully established. The other is to integrate trees and shrubs permanently into the crop or animal production system, to the benefit of both crop production and land resource protection. Thus, agroforestry encompasses many traditional land-use systems such as home gardens, shifting cultivation and bush fallow systems.

Alley cropping can be considered an improved bush fallow system. Small trees or shrubs are planted in cropland in rows, preferably along the contour (even where east-west orientation of the rows may minimize shading of crops). The optimal spacing between rows depends on: slope; soil type and its susceptibility to erosion; rainfall; crop species; and the soil and crop management system (FAO, 1995).

Besides adding organic matter to the system, perennial trees and shrubs recycle plant nutrients from deeper soil layers through their rooting system. Through litter and pruning, these can be used again by annual crops. Probably the most important contribution of perennials in a production system lies in the fact that throughout the whole year their roots excrete root exudates and decaying root cells, which in turn are used as an energy source by soil microorganisms. The food web in the soil is maintained, even during dry seasons when no annual crops are grown. The result is that soil biota are in place to provide the crop with nutrients at the beginning of the next cropping season.

Direct seeding is the easiest and cheapest way of establishing hedgerows around fields or in the fields (alleys). However, emerging seedlings may not be able to compete with weeds without additional care. Therefore, starting plant growth in a nursery and transplanting may be necessary for some species. Other species may be established by cuttings. With good establishment, the plants will be better able to withstand both dry spells and browsing by livestock. Crucial to a successful establishment of the hedgerow is that the selected plants should be tall enough to outgrow the weeds at the time of the first crop harvest.

During the cropping season, hedgerow pruning is needed in order to avoid shading of the crop (Plate 11). The timing, frequency and extent of pruning depend on the species used and the season. As a general rule, the lower the hedgerows and the taller the crop, the less frequently is pruning required. Fastgrowing plants such as Leucaena leucocephala and Gliricidia sepium may require pruning every six weeks during the cropping season. They are often pruned to a height of about 50 cm. Care must be exercised as too frequent pruning can result in tree dieback.

Figure: *Agroforestry plot with suboptimal stand of maize because of shading by trees.*

The integration of trees and woody shrubs into the cropping system offers additional uses and many benefits, as mentioned by farmers using the Quezungual system in Honduras (Plate 12 and Box 6). However, farmers with short-term land tenure may not be interested in these benefits. Furthermore, the plantation of trees sometimes has an effect on the land tenure status; therefore tenants may not be allowed to establish trees on agricultural land. Agroforestry systems can also inhibit mechanization and may need increased labour inputs, especially for hedgerow pruning.

Figure: *Example of the "Quezungual" system, an indigenous agroforestry system.*

The hedgerow species have to be selected carefully in order to avoid negative impacts on crop production because of the complex relationships (competition for light, water and nutrients, allelopathy, occurrence of pest and diseases, etc.) that are inherent to agroforestry systems. Many farmers may consider the hedgerows as not useful, especially where their positive effects are not secure or visible. Where livestock are allowed to graze freely, it can be difficult to establish hedgerows without taking special measures to protect the young plants. Fencing or control of grazing animals may require collective efforts and agreement by the local community. The increased labour requirement, the reduced cropped area and the difficulty of mechanization may make alley cropping uneconomic unless the hedgerow species produce direct benefits such as fruits, fuelwood or poles/timber for construction purposes (in addition to the nutrient recycling and erosion control effects).

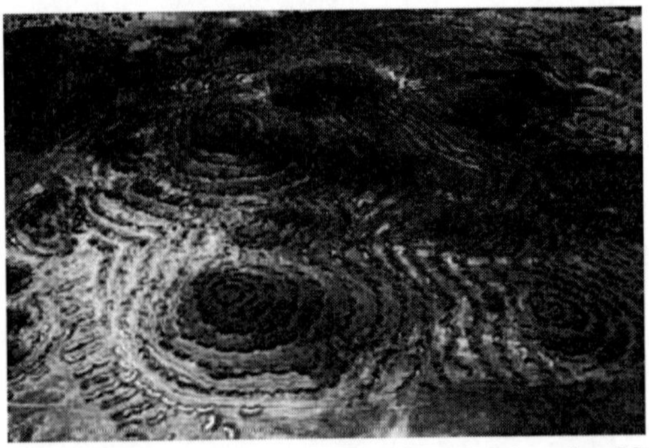

Figure: *Afforestation of "badlands" in the Niger.*

Reforestation and afforestation: Afforestation means the establishment of a forest on land that has not grown trees recently. It can serve two principal soil and water conservation purposes: protection of erosion-prone areas, and revegetation and rehabilitation of degraded land. Afforestation is specifically used to provide protective cover in vulnerable, steep and mountainous areas. Afforestation helps to replenish timber resources and provide fuelwood and fodder.

The establishment of a forest cover under good management is an effective means of increasing organic matter production. However, the land must have the productive capacity to support an appropriate forest type, which differs according to climate, soil, slope and the specific purpose of the forest (timber production, livestock grazing, etc.). Therefore, the choice of species and the selection of an appropriate site are of particular importance for successful afforestation.

The procurement of adequate quantities of good quality seed of the species and provenances (adapted varieties) required is a prerequisite for any afforestation effort. However, it is often difficult to find suitable and reliable sources of such seeds.

A number of species require special pre-treatment of the seed or seedling in order to achieve satisfactory germination and uniform stand. Such treatment may consist of soaking the seed in water for varying lengths of time, alternate soaking and drying, scarifying or chipping the seed coat to render it permeable to water, plunging the seed into boiling water or even boiling it for a short time. Some tree seedlings may need a mycorrhizal treatment when planted in soils that are deprived of associated mycorrhizae species as well as rhizobium species (e.g. Casuarina in Senegalese sandy soils). The aim is to ensure that good numbers of plants germinate and that germination after sowing is both rapid and uniform.

Afforestation can be achieved by direct sowing or replanting young plants from a nursery. The main advantage of direct sowing is the reduced cost. However, this is usually much less reliable and is only justified where:

- seed is plentiful and cheap;
- adequate germination under field conditions can be relied on;
- the seedlings send down a deep tap root rapidly and are able to withstand adverse climatic conditions in the time after germination;
- the rate of growth is sufficiently fast to make a prolonged period of tending and weeding unnecessary.

Regeneration of natural vegetation

Regeneration of natural grasslands and forest areas increases biomass production and improves the plant species diversity, resulting in more diverse soil biota and other associated beneficial organisms. Natural regeneration may be more reliable where land is not very productive. In some cases, natural regeneration of a given area may lead to the infestation of plots by weeds. Increasingly, natural vegetation is being recognized for its multipurpose benefits, for example, fuelwood, fibre, biocontrol (e.g. neem) and medicinal species, as well as restoration of soil fertility (Acacia albida and other leguminous species) and habitats for various beneficial species (pollinators and natural enemies) as well as wildlife.

Increased Organic Matter Supply

Protection From Fire

Burning affects organic matter recycling significantly. Fire destroys almost all organic materials on the land surface except for tree trunks and large branches. In addition, the surface soil is sterilized, loses part of its organic matter, the population of soil microfauna and macrofauna is reduced, and no ready-to-use organic matter is available for rapid restoration of the populations. However, this practice is widely used (e.g. in Africa) in order to enhance pasture regrowth for livestock (using residual P), to control pests and diseases, and even to catch small animals for food.

A specific and difficult case is the burning of sugar cane before harvesting. It has both a technical dimension (CO2 and greenhouse gas emissions, mechanization of harvest, sugar content, etc.) and a social dimension (manual cutting, source of survival resources for poor/landless workers). The damage depends on fire intensity, which is a function of vegetation type and climate conditions and frequency. The costs and benefits of burning and the methods to minimize harmful effects need to be identified with local populations.

Crop Residue Management

In systems where crop residues are managed well, they:

- add soil organic matter, which improves the quality of the seedbed and increases the water infiltration and retention capacity of the soil, buffers the pH and facilitates the availability of nutrients;
- sequester (store) C in the soil;
- provide nutrients for soil biological activity and plant uptake;
- capture the rainfall on the surface and thus increase infiltration and the soil moisture content;
- provide a cover to protect the soil from being eroded;
- reduce evaporation and avoid desiccation from the soil surface.

Depending on the nature of the following crop, decisions are made as to whether the residues should be distributed evenly over the field or left intact, e.g. where climbing cover crops (e.g. mucuna) use the maize stalks as a trellis.

An even distribution of residues: (i) provides homogenous temperature and humidity conditions at sowing time; (ii) facilitates even sowing, germination and emergence; (iii) minimizes the development of pests and diseases; and (iv) reduces the emergence of weeds through allelopathic effects. The most appropriate method for managing crop residues depends on the purpose of the crop residues and the experience and equipment available to the farmer. Where the aim is to maintain a mulch over the soil for as long as possible, the biomass is best managed using a knife roller, chain or sledge in order to break it down but not kill it. Where the decomposition process should commence immediately in order to release nutrients, the residues should be slashed or mown and some N applied because dry residues have a high C:N ratio. However, in order to avoid nitrate emission, urea should not be broadcast on the surface but injected where possible.

Figure: *Crop residue management by using a knife roller.*

Utilizing forage by grazing rather than by harvesting: In many places, there is competition for the use of crop residues that can be used as fodder, for roofing, artisan handicrafts, etc. Where residues are to be used for animal feed, either the animals graze the residues directly, or they are stall- or kraal-fed.

Removal of the residues from the field can lead to a considerable loss of organic matter where animal manure is not returned to the field. By controlled grazing, the animal manure is returned in the field without a high labour input. The experience of Guaymango, El Salvador, demonstrates that it is possible to achieve successful integration of crop and livestock components without creating competition in the allocation of crop residues (Vieira and Van Wambeke, 2002). The amount of residues produced by the system is enough to serve both as soil cover and as fodder for livestock, mainly because of the use of local sorghum varieties (instead of HYVs) that have a high straw/grain ratio. As farmers value crop residues as soil cover, a fodder market has developed where grazing rights, number of cattle and duration of grazing are traded. In the northern zone of the United Republic of Tanzania, farmers have found a compromise between using the residues for grazing or soil cover, albeit one that is rather labour intensive. They separate the palatable and non-palatable parts of the crop residues. They use the non-palatable parts to cover the soil and act as food for soil organisms, while they feed the palatable parts to cattle and goats that are kept close to the homestead.

Integrated pest management: As with balanced fertilization, proper pest and disease management results in healthy crops. Healthy crops produce optimal biomass, which is necessary for organic matter production in the soil. Diversified cropping and mixed crop-livestock systems enhance biological control of pests and diseases through species interactions. Through integrated production and pest management farmers learn how to maintain a healthy environment for their crops. They learn to examine their crops regularly in order to observe ratios of pests to natural enemies (beneficial predators) and cases of damage, and on that basis to make decisions as to whether it is necessary to use natural treatments (using local products such as neem or tobacco) or chemical treatments and the required applications.

Figure: *Farmer giving fodder to cattle*

Applying animal manure or other carbon-rich wastes: Any application of animal manure, slurry or other carbon-rich wastes, such as coffee-berry pulp, improves the organic matter content of the soil. In some cases, it is better to allow a period of decomposition before application to the field. Any addition of carbon-rich compounds immobilizes available N in the soil temporarily, as micro-organisms need both C and N for their growth and development. Animal manure is usually rich in N, so N immobilization is minimal. Where straw makes up part of the manure, a decomposition period avoids N immobilization in the field.

Compost: Composting is a technology for recycling organic materials in order to achieve enhanced agricultural production. Biological and chemical processes accelerate the rate of decomposition and transform organic materials into a more stable humus form for application to the soil. Composting proceeds under controlled conditions in compost heaps and pits (Müller-Sämann, 1986).

Compost heaps should have a minimum size of 1 m and are suitable for more humid environments where there is potential for watering the compost. Compost pits (Plate 16) should be no deeper than 70 cm and should be underlain with rough material for good aeration of the compost. Pits are suitable for drier environments where the compost may desiccate (Müller-Sämann, 1986). Dry composting relies on covering the compost with soil and creating an anaerobic environment. However, this is a slower process than the more usual moist aerobic process. The ratio of C to N in the compost pile is important for optimizing microbial activity. Thus, a mixture of soft, green and brown, tougher material is used. Ash and phosphate rock are often added to accelerate the process.

Composting can complement certain crop rotations and agroforestry systems. It can be used efficiently in planting pits and nurseries. It is very similar in composition to soil organic matter. It breaks down slowly in the soil and is very good at improving the physical condition of the soil (whereas manure and sludge may break down fairly quickly, releasing a flush of nutrients for plant growth). In many circumstances, it takes time to rejuvenate a poor soil using these practices because the amount of organic material being added is small relative to the mineral proportion of the soil.

Successful composting depends upon the sufficient availability of organic materials, water, manure and "cheap" labour. Where these inputs are guaranteed, composting can be an important method of sustainable and productive agriculture. It has ameliorative effects on soil fertility and physical, chemical and biological soil properties. Well-made compost contains all the nutrients needed by plants. It can be used to maintain and improve soil fertility as well as to regenerate degraded soil. However, materials for compost production may be in short supply and the technology demands high labour inputs for proper compost production and application. Therefore, compost application may be restricted to certain crops and limited application areas, e.g. vegetable production in home gardens.

Mulch or permanent soil cover: One way to improve the condition of the soil is to mulch the area requiring amelioration. Mulches are materials placed on the soil surface to protect it against raindrop impact and erosion, and to enhance its fertility (FAO, 1995). Crop residue mulching is a system of maintaining a protective cover of vegetative residues such as straw, maize stalks, palm fronds and stubble on the soil surface (Plate 17).

The system is particularly valuable where a satisfactory plant cover cannot be established rapidly when erosion risk is greatest (FAO, 1993).

Mulching adds organic matter to the soil, reduces weed growth, and virtually eliminates erosion during the period when the ground is covered with mulch.

There are two principal mulching systems:

- in situ mulching systems - plant residues remain where they fall on the ground (Plate 17);
- cut-and-carry mulching systems - plant residues are brought from elsewhere and used as mulch (Plates 18 and 19).

Crop residue mulching has numerous positive effects on crop production. However, it may require a change in existing cropping practices. For example, farmers may conventionally burn crop residues instead of returning them to the soil. In situ mulching depends on the design of appropriate cropping systems and crop rotations, which have to be integrated with the farming system. The greater labour demands of cut-and-carry systems represent a major constraint. Mulch may be more relevant in home gardens or for valuable horticulture crops than in less intensive farming systems.

Figure: *No-till maize under black oat mulch, Santa Catarina, Brazil*

Mulch affects the soil life. Holland and Coleman (1987) have demonstrated that litter placement on the soil surface (as opposed to incorporation with ploughing) increased the ratio of fungi to bacteria

- the reason being that fungi have a higher carbon assimilation efficiency than bacteria. In addition, it encourages bioturbating (mixing) effects of macrofauna that pull the materials into surface layers of the soil.

Decreased Decomposition Rates

Reduced or zero tillage

Repetitive tillage degrades the soil structure and its potential to hold moisture, reduces the amount of organic matter in the soil, breaks up aggregates, and reduces the population of soil fauna such as earthworms that contribute to nutrient cycling and soil structure.

Avoiding mechanical soil disturbance implies growing crops without mechanical seedbed preparation or soil disturbance since the harvest of the previous crop. The term zero tillage is used for this practice synonymously with terms such as no-till farming, no tillage, direct drilling, and direct seeding.

Compared with conventional tillage, reduced or zero tillage has two advantages with respect to soil organic matter. Conventional tillage stimulates the heterotrophic microbiological activity through soil aeration, resulting in increased mineralization rate. Through breakdown of soil structure, it decreases upward and downward movements of soil fauna, such as earthworms, which are largely responsible for "humus" production through the ingestion of fresh residues. Reduced or zero tillage regulates heterotrophic microbiological activity because the pore atmosphere is richer in $CO2/O2$, and facilitates the activity of the "humifiers" (Pieri et al., 2002).

***Figure:** Farmer at work in a field of lettuces.*

Tillage has become the most common method to control weeds. However, mulching is a more environmentally sound practice than tillage for weed control. The loose soil that results from tilling has less structure than before; the appearance is deceptive. Subsequent traffic or heavy rain soon packs this loosened soil, not only negating the expensive cultivation that produced the loose soil but also culminating in a degraded environment for water entry, seed germination and root growth. Further cultivation is then required to re-loosen the soil; more expense with the same outcome - subsequent repacking and degraded soil structure.

This is a typical "downward spiral" of conventional agriculture. Moreover, tillage when the soil is too moist or too dry leads to compaction or pulverization of soil; but farmers may not have the option to wait for optimal conditions. Severe, accelerated soil erosion and the high costs in terms of labour and energy associated with plough-based methods of seedbed preparation have led to the widespread adoption of no- or zero-tillage systems for cropping in temperate and tropical climates.

In no-tillage systems, the crop is sown into a soil left undisturbed since the harvest of the previous crop. Crop residue mulch is maintained and anchored firmly to the ground. Weed control relies on mechanical slashing or cover crops (FAO, 1993). Contact herbicides are also used in some cases. In reduced- or zero-tillage systems, soil fauna resume their bioturbating activities gradually. These loosen the soil and mix the soil components (also known as biotillage). The additional benefit of the increased soil organic matter and burrowing is the creation of a stable and porous soil structure without expensive, time-consuming and potentially degrading cultivations. In zero-tillage systems, the action of soil macrofauna gradually incorporate cover crop and weed residues from the soil surface down into the soil. The activity of microorganisms is also regulated by the activity of the macrofauna, which provide them with food and air through their burrows. In this way, nutrients are released slowly and can provide the following crop with nutrients.

Several authors have demonstrated that some crop rotations and zero tillage favour Bradyrhizobia populations, nodulation and thus N fixation and yield (Voss and Sidirias, 1985, Hungria et al., 1997, Ferreira et al., 2000). Figure 10 indicates a 200-300-percent increase in population size of root nodule bacteria in a zero-tillage system compared with conventional tillage. The presence of soybean in the crop rotation resulted in a fivefold to tenfold increase in population size of the same bacteria compared with cropping systems without soybean.

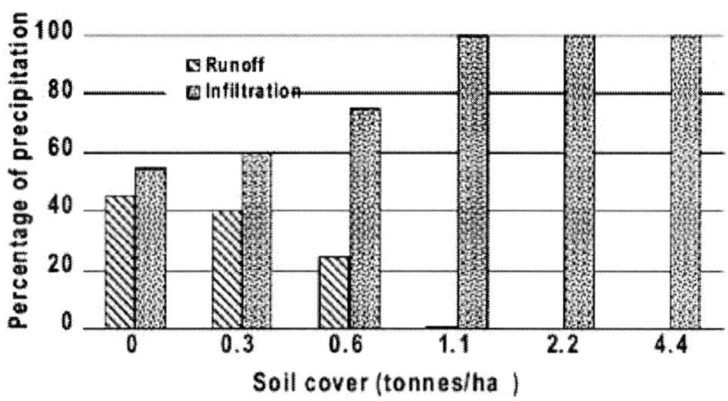

Figure: *Population size of root nodule bacteria with different crop rotations*

Note: S = soybean; W = wheat; M = maize.

Source: Voss and Sidirias, 1985.

Strictly speaking, the term zero tillage applies to methods involving no soil disturbance whatsoever, a condition that may be difficult to achieve. Broadcasting of seed is one way of applying zero tillage (Plate 20). The seed is broadcast over the previous crop residues and, where necessary, the residues are shaken to ensure that the seed falls on the soil surface.

In direct drilling, seeds such as maize, sorghum, soybean, wheat and barley are sown directly into shallow furrows cut into the previous crop residues. Weeds are controlled mechanically with a knife, which knocks down the plants and breaks their stems, or chemically with herbicides.

Traditional practices such as the burning of crop residues may inhibit the introduction of no-tillage systems. In many situations, a conflict exists between leaving crop residues on the surface or feeding them to livestock in the dry season when there is a shortage of fodder.

Mechanical soil disturbance also includes soil compaction through wheel impact of machinery, especially important in large-scale mechanized agriculture, e.g. plantations (sugar cane) or biannual crops (cotton). In a zero-tillage farming system, consideration must be given to reducing both the random placement of tyres/wheels in fields as well as the potential for compaction from animal hooves. Pietola, Horn and Yli-Halla (2003) reported the destructive effect of cattle trampling on the soil structure. Proffitt, Bendotti and McGarry (1995) demonstrated the almost total loss of soil porosity in the soil surface as a result of trampling by sheep. There is a belief that draught animals cause less land degradation than tractors. However, there are reports of soil compaction on smallholder farming enterprises in both Malawi (Douglas et al., 1999) and Bangladesh (Brammer, 2000). The hooves of draught animals and the shearing effect of ploughs or hand hoes, which are used repeatedly at a constant depth, can cause severe compacted layers. Grazing animals should be removed from zero-till fields in moist-wet soil conditions as the compaction risk is greatest at these times.

Controlled traffic, where the wheels of all in-field equipment follow permanent, defined tracks, ensures that compaction is restricted to specific known areas (Plate 22). Alternatively, flotation tyres (low ground-pressure tyres) should be fitted to all large tractors, harvesters, in-field grain bins, etc. in order to reduce their compacting potential.

Recent research has demonstrated the devastating effects of compaction from wheel impact on the occurrence and survival of eartworms (Pangnakorn et al., 2003). Earthworm incidence was greater under controlled traffic than under wheeled traffic. Figure 11 shows the immediate effects of wheeling and tillage on the earthworm population. It appears that wheeling has the most detrimental effect on earthworm survival and that where wheeling is followed by tillage the survival rate is much greater. This may indicate that earthworms are able to survive an initial compaction in the field as long as it is relieved immediately. Where it is not, earthworms are inmobilized and unable to find air and nutrients.

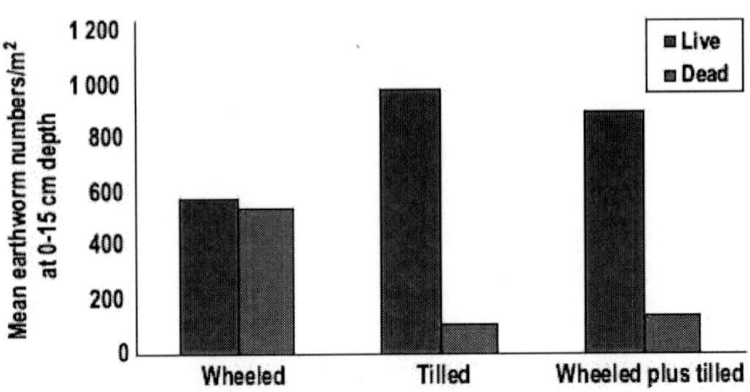

Figure: *Live and dead earthworm numbers per square metre at 0-15 cm of soil depth sampled immediately after treatment*

Organic Matter Deposition

The reduction of soil disturbance through zero-tillage, the use of cover crops and the preservation of crop residues on the soil surface result in increased activity of the soil and in the accumulation of organic matter, mainly in the topsoil.

An argument often heard in the discussion on conservation agriculture is that it is only feasible in the humid and subhumid tropics and that the generation of sufficient biomass in semi-arid regions is the limiting factor to start implementing conservation agriculture.

However, recent research has shown that even in semi-arid areas of Morocco the application of the principles of conservation agriculture bears its fruits. Mrabet (2000) reports higher yields through better water use and improved soil quality; the latter caused by an increase in soil organic C and N and a slight pH decline in the seedzone.

Increased Carbon Sequestration

World soils are important reservoirs of active C and play a major role in the global carbon cycle. As such, soil can be either a source or sink for atmospheric CO2 depending on land use and the management of soil and vegetation (Lal, 2005). The conversion of native ecosystems (e.g. forests, grasslands and wetlands) to agricultural uses, and the continuous harvesting of plant materials, has led to significant losses of plant biomass and C, thereby increasing the CO2 level in the atmosphere. In particular, the practice of burning agricultural fields before cultivation has a disastrous effect on soil organic carbon content.

Chapter 3

Nitrogen: Role, Dynamics and Management

Introduction to Nitrogen

Nitrogen is one of the macronutrients which is required in large amount for plant metabolism and growth act as a primary nutrient for plants. It is absorbed in Ammonium (NH4+), Nitrate (NO3-) ions forms. Nitrogen is the element which is not directly available to plants from atmosphere and earth's crust.

Nitrogen Forms and Function

Forms of nitrogen available for plant uptake

- Ammonium
- Nitrate

Functions of nitrogen in plants

- Nitrogen is an essential element of all amino acids. Amino acids are the building blocks of proteins.
- Nitrogen is also a component of nucleic acids, which form the DNA of all living things and holds the genetic code.
- Nitrogen is a component of chlorophyll, which is the site of carbohydrate formation (photosynthesis). Chlorophyll is also the substance that gives plants their green colour.
  - o Photosynthesis occurs at high rates when there is sufficient nitrogen.
  - o A plant receiving sufficient nitrogen will typically exhibit vigorous plant growth. Leaves will also develop a dark green colour.

Importance of Nitrogen in Soil

- In soil atmospheric nitrogen is a major source of nitrogen. In the atmosphere, it exists in N- form and must be converted before it becomes useful in the soil.
- Rhizobia is a bacterium that infects the roots of and receives much food energy from, legume plants can fix much more nitrogen per year. The quantity of nitrogen fixed by Rhizobium bacteria exceeds that needed by the microbes themselves for use by the host legume plant.
- Nitrogen has three forms as soil organic nitrogen compounds, ammonium (NH) ions and nitrate (NO) ions. In soil, nitrogen is available in organic forms in a natural way as plant and animal residues but this nitrogen is not directly available to plants it needs microorganism to convert so that they can available for plants. The nitrogen in soil minerals is generally slow and contributes only slightly to nitrogen in the soil. The available nitrogen is in the inorganic forms NH-- and NO-- these forms are mostly available in the soil. Ammonium ions bind to the soil in negatively charged cation exchange complex (CEC). Nitrate ions do not bind to the soil solids because they have negative charges, but exist dissolved in the soil water, or precipitated as soluble salts under dry conditions.

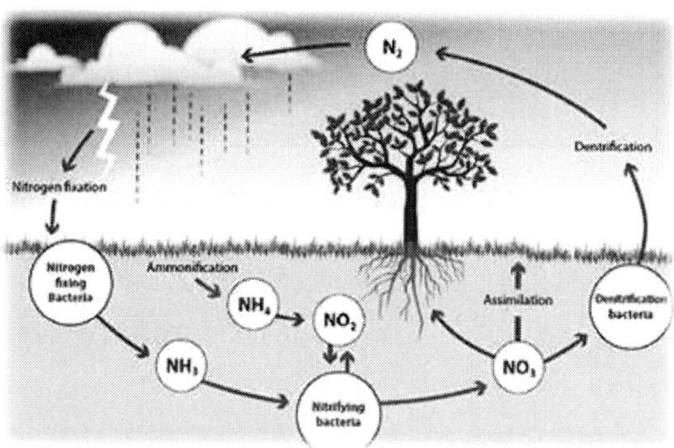

Figure: Nitrogen in Soil

Importance of Nitrogen in Plants

- Nitrogen is a major component of chlorophyll, the compound by which plants use sunlight energy for photosynthesis to produce sugars from water and carbon dioxide.
- It plays an important in soil fertility management so it has higher concentration compared to carbon, hydrogen and oxygen.

- Nitrogen imparts vigorous vegetative growth dark green colour to plants.
- Nitrogen is an energy-transfer compound, such as ATP (adenosine triphosphate). ATP allows cells to conserve and use the energy released in metabolism in plants.
- Nitrogen is a major component of amino acids.
- It is an essential constituent of protein and present in many other compounds of physiological importance in plant metabolism.
- Nitrogen produces early growth and also results in delay in maturity in plants.
- A complex group of proteins such as Nucleoproteins involved in the development and hereditary process. It is a significant component of nucleic acids such as DNA (Deoxyribonucleic acid), RNA (Ribonucleic acid) the genetic material that allows cells to grow and reproduce.
- The chief source of Nitrogen in the soil because plant absorb it either in the form of nitrate or Ammoniacal salt and it governs the utilization of potassium, phosphorus and other elements.
- When the temperature is lower than higher is the N content due to more organic matter addition and slow rate of decomposition.
- Improves the quality of leafy vegetables and fodders and also increasing protein content Some bacteria and heterocyst containing blue-green algae fix nitrogen of the atmosphere which can be utilised by plants.
- Ammonium ($NH4+$) fixation increases with increase in 15 the content particularly 2:1 type of clay minerals like vermiculite, fine-grained mica and smectite.
- Fixation of ammonium ion ($NH4+$) is generally higher in subsoil due to higher clay content and a lower rate of nitrification.

Nitrogen Cycle

Plants need nitrogen to grow, develop and produce seed. The main source of nitrogen in soils is from organic matter. Soils in Missouri commonly contain one to four percent organic matter. Organic matter largely arises from plant and animal residues. The nitrogen in organic matter is largely in organic forms that plants cannot use. Bacteria found in soils convert organic forms of nitrogen to inorganic forms that the plant can use. Nitrogen is taken up by plant roots and combined into organic substances in the plant, such as enzymes, proteins and chlorophyll. Chlorophyll gives the plant its green colour. When the plant dies, it decays and becomes part of the organic matter pool in the soil.

The nitrogen cycle is a biological process through which nitrogen is converted into ammonium ion and nitrates ions by the atmosphere to soil to the organism and again into the ecosystem.

It exists in two forms organic forms and inorganic forms. Organic forms involve animal waste, manures, residue etc they all go through the decomposition process. Inorganic forms involve symbiotic bacteria which converts nitrogen into the usable form so that plants can absorb.

It has several processes such as Nitrogen fixation, Nitrification, Assimilation, Ammonification and Denitrification.

- Plant and animal wastes decompose, adding nitrogen to the soil.
- Bacteria in the soil convert those forms of nitrogen into forms plants can use.
- Plants use the nitrogen in the soil to grow.
- People and animals eat the plants; then animal and plant residues return nitrogen to the soil again, completing the cycle.

Another way nitrogen enters the cycle is as inorganic nitrogen from the atmosphere and factories. The concern with these forms is that the incremental amount of nitrates they add to the nitrogen cycle may threaten groundwater.

- Rain storms contribute atmospheric nitrogen through rain drops that reach the soil.
- Legumes, such as soybeans, alfalfa and clovers, are plants that can convert atmospheric nitrogen into plant-usable nitrogen.
- Factories that produce nitrogen fertilizers add nitrogen to the soil when farmers and gardeners "feed" their crops.
- Nitrogen in sewage sludge from municipal waste plants can be used to fertilize farm fields.

Nitrogen Fixation

- The first step in the Nitrogen cycle.

- It converts Atmospheric nitrogen (N_2) which is inert form into usable ammonia (NH_3).
- Atmospheric nitrogen source is from the atmosphere, surface waters through precipitation, rainfall.
- It undergoes changes into two nitrogen atoms and combines with hydrogen to form ammonia.
- Symbiotic bacteria involved in this process is Diazotrophs and azotobacter, rhizobium also has a major role in this process.

Types of Nitrogen fixation involved are

- Atmosphere fixation: Natural phenomenon (energy of lightning)
- Industrial nitrogen fixation: Man-made alternative (Urea)
- Biological nitrogen fixation: Bacteria like Rhizobium and blue-green algae fixes nitrogen components in soil.

Nitrification

- The ammonia is converted into nitrate by oxidation of ammonia with help of Nitrosomonas bacteria and later produce nitrites into nitrates by Nitrobacter.
- Nitrification reaction:

$$2NH^{4+} + 3O_2^- \, 2NO_2^- + 4H^+ + 2H_2O$$

$$2NO_2^- + O_2 - 2NO_3^-$$

Assimilation: The process of converting nitrogen compounds such as ammonium ion, nitrate ion, nitrite ion is used for the formation of plants and animal protein.

Ammonification

- When organic material decomposes the nitrogen present in these matters are released into the soil and converts that material into ammonium.
- These are decomposed by bacteria or fungi in soil and produces ammonia which is used for biological processes.

Denitrification

- Denitrification is the final step in the Nitrogen cycle.
- The process in which nitrate (NO3-) converts into Nitrogen(N) is called Denitrification and in this process, nitrate to gain oxygen and gives out free nitrogen as a by-product.
- Bacteria involved in this process are Clostridium and pseudomonas bacteria.

Ways nitrogen is lost to the cycle

For the most part, the nitrogen cycle is soil based. Nitrogen is lost from the cycle in four ways:

- Denitrification: Bacteria change nitrate in the soil to atmospheric nitrogen, which joins the atmosphere.
- Volatilization: Turns urea fertilizers and manures on the soil surface into gases that also join the atmosphere.

  Together, these first two processes account for most of the nitrogen lost to the cycle - a concern for soil fertility.
- Runoff: Carries the nitrogen in fertilizers and manure and the nitrogen in the soil into our rivers and streams - a concern for water quality.
- Leaching: Carries nitrates soo deep into the soil that plants can no longer use them, producing a dual concern - for lost fertility and for water quality, as nitrates enter the groundwater and the wells that provide our drinking water.

More About the Nitrogen Cycle

The largest single source of nitrogen is the atmosphere. It is made up of 78 percent of this colourless, odorless, nontoxic gas.

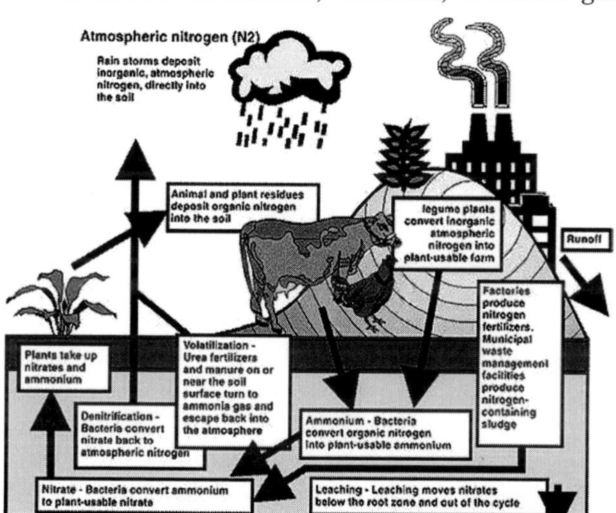

Figure: *A more comprehensive look at the nitrogen cycle.*

However, plants are unable to use nitrogen as it exists in the atmosphere. Nitrogen from the air (N2) enters the nitrogen cycle through several unique types of microorganisms that can convert N2 gas to inorganic forms usable by plants. Some of these microorganisms live in the soil, while others live in nodules of roots of certain plants.Nitrogen also can enter the cycle from other sources besides the air, manure and decaying plant materials.

Nitrogen also can enter the cycle from the application of commercial nitrogen fertilizers. Nitrogen can be lost from the cycle. It can be lost to the atmosphere, removed by harvesting crops or lost to surface water or groundwater. However it is lost, nitrogen can enter the cycle again through one of the processes discussed above or through other processes. These additional pathways of gains and losses to the nitrogen cycle are illustrated in figure.

Impact on Water Quality

Nitrogen becomes a concern to water quality when nitrogen in the soil is converted to the nitrate (NO3) form. It is a concern because nitrate is very mobile and easily moves with water in the soil. The concern of nitrates and water quality is generally directed at groundwater.

However, nitrates can also enter surface waters such as ponds, streams and rivers. The presence of nitrates in the soil are largely the result of natural biological processes associated with the decomposition of plant residues and organic matter. Nitrates can also come from rainfall, animal manure and nitrogen fertilizers.Whether or not nitrates actually enter groundwater depends on underlying soil and/or bedrock conditions, as well as the depth to groundwater.

If depth to groundwater is shallow and the underlying soil is sandy, the potential for nitrates to enter groundwater is relatively high. However, if depth to groundwater is deep and the underlying soil is heavy clay, groundwater contamination from nitrates is not likely.

Once nitrates get into the groundwater, the greatest concerns are for infants less than one year old and for young or pregnant animals. High levels of nitrates can be toxic to newborns, causing anoxia, or internal suffocation. Seek alternative water sources if nitrate levels exceed the health standard of 10 ppm nitrate-N. Do not boil water to eliminate nitrates. It increases nitrate levels rather than decreasing them. The most common symptom of nitrate poisoning in babies is a bluish colour to the skin, particularly around the baby's eyes and mouth. These symptoms of nitrate toxicity are commonly referred to as the "blue-baby" syndrome.

Nitrogen Fixation

Although atmospheric nitrogen gas (N2) makes up approximately 78% of the air, it cannot be directly used by plants. Instead, atmospheric N2 only becomes available to plants through three unique processes. The final product of each of these processes is ammonium, which is then available for plant uptake. The three processes which convert atmospheric nitrogen to ammonium

- biological nitrogen fixation
- chemical nitrogen fixation
- atmospheric addition

Biological nitrogen fixation

Certain soil organisms have the special ability to convert atmospheric nitrogen to ammonium. These organisms include several species of bacteria, actinomycetes, and cyanobacteria. In the soil, nitrogen fixating organisms can form special relationships with plants, called "symbiotic" associations. Symbiotic is a term that means "living together." Although a symbiotic relationship can be antagonistic, the symbiosis that occurs during biological nitrogen fixation is generally mutual and beneficial.

Legume-Rhizobium Symbiosis: The most abundant symbiotic relationship in nitrogen fixation forms between legumes (i.e. alfalfa, soybeans, etc.) and the Rhizobia bacteria species.

- As the roots of legumes grow, Rhizobium bacteria infect the root hairs where they begin to multiply.
- As a response to this colonization, the legume forms nodules, which are structures that form around the Rhizobia.
- Within these nodules, Rhizobia bacteria are able to continue multiplying and converting the N2 from the soil air to ammonium.
    - o However, the presence of nodules is not a sufficient indicator that nitrogen is being converted to ammonium. Active and effective nodules are generally greater than 2 mm, have pink to red interiors, and concentrate around the tap root.
    - o On the other hand, non-effective nodules are generally smaller in size with white, green or brown interiors.The image below shows cross sections of soybean (Glycine max) nodules. Nodules in the first row are highly effective at converting N2 from the atmosphere into ammonium. The

nodules in the second row are moderately effective at biological nitrogen fixation, while the bottom row of nodules do not fix any nitrogen at all. Notice the lack of colour in the interior of these nodules, which indicates that they do not have an active 'nitrogenase system'. A 'nitrogenase system' is the bacterial enzyme that is necessary to convert N2 gas into ammonium through this biological process.

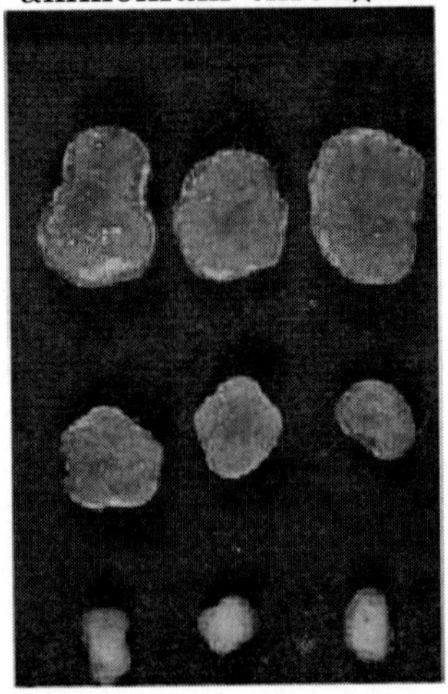

Figure: *Relationship between colour and effectivity of nodulation in soybean.*

Source: Legumes Inoculants and Their Use, 1984. University of Hawaii NifTAL Project and FAO.

The next image shows how a well nodulated soybean root system looks when it contains many highly effective rhizobia in the soil or is applied as an inoculant.

Specificity: Some Rhizobium species are only capable of nodulating a particular legume species and cannot successfully nodulate other legumes.

- For example, the Rhizobium that nodulates alfalfa is a different species from the Rhizobium that nodulates soybean.

This phenomenon is known as Rhizobium specificity. However, not all Rhizobium are legume-specific. Thus, some may nodulate a number of different legumes. The figure below gives examples of some common cross-inoculation groups that may assist you in selecting the proper rhizobial inoculant for a particular legume host. The proper combination of rhizobia and legume will result in the optimal nodulation and most nitrogen fixation. From this figure, we see that using soybean rhizobia with a soybean plant forms an effective symbiosis, while using a soybean rhizobia with a leucaena plant does not. However, cowpea rhizobia is capable of nodulating both mungbean and peanut.

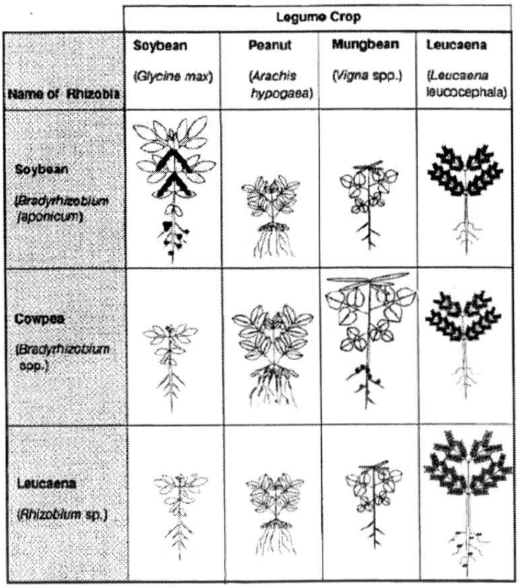

Figure: *Specificity of Rhizobia for successful nodulation of certain legumes.*

Biological Nitrogen Fixation Management Program: The occurrence of the symbiotic relationship is heavily dependent upon a variety of soil conditions. If your program incorporates nitrogen fixation, the following considerations can determine your success.

- First and foremost, the Rhizobium must be compatible with the legume. If your crop is Rhizobium specific, you must use the correct Rhizobium species.
- If your inoculum (which contains the Rhizobium bacteria) is applied to seeds, the procedures must be properly followed.
- Nitrogen fixation takes place when total soil nitrogen is insufficient. When sufficiently present, the plant will instead rely on the nitrogen available from the soil.
- Rhizobia are sensitive to any growth factor that limits root development. Such conditions as aluminum and manganese toxicities will limit inoculation.
- Rhizobia are influenced by mineral nutrient imbalances.
    - o Low levels of calcium, phosphate, molybdenum under acidic conditions will limit nitrogen fixation.

- o Under alkaline conditions, phosphate, cobalt, boron, iron, and copper levels become a concern.
- Any growth factor (such as light, water temperature stresses or soil compaction) and any management factor (such as nutrient management, salinity) that detrimentally affects growth of the legume will detrimentally impact nitrogen fixation.

Amount of Nitrogen Fixed by Legume: When nitrogen is converted to ammonium during biological nitrogen fixation, ammonium becomes available to the legume and the microorganism that fixes it.

Typically, the bacteria can fix anywhere between 20 and 80% of the total legume N.

- Perennial legumes may fix 100 to 200 lb/a/yr
- Annual legumes fix 50 to 100 lb/a/yr.

Small amounts of ammonium can also be released by roots of the legume into the rhizosphere, or the surrounding soil.

Availability of Nitrogen in Subsequent Cropping Systems

What about subsequent cropping systems- Can a legume rotation benefit later plantings of a nonlegume crop-

Research shows that yields of nonlegume crops can increase when following a legume rotation. It is believed that the legume rotation increases the N content of soil, thus making it an effective nutrient management strategy.

However, when the legume is incorporated into the soil, the major benefit of the legume rotation lasts only during the first year following the legume rotation.

Other symbiotic relationships: Legumes and Rhizobia are not the only species that can establish a mutual symbiotic relationship needed for nitrogen fixation to oocur.

- In wetland rice production, a symbiotic relationship may form between Anabaena azolla (a blue green algae) and the Azolla fern. As a result, wetland plants can benefit by incorporations of Azolla as a green manure.
- Certain tree species (i.e Causarina sp.) can form symbiotic relationships with certain species of Actinomycetes and the Frankia bacteria. Though this relationship has lesser agricultural importance, it may gain significance in forestry, or wood production.

Free-Living Nitrogen Fixation: "Free-living" nitrogen fixating organisms are also capable of nitrogen fixation, but are not associated with any plant species.

- Examples of these organisms are azotobacteria, azolospirillum, and clostridium. However, free-living species do not contribute largely to agricultural production

Chemical Nitrogen Fixation

Since the 1950s, ammonium-based fertilizers have been manufactured using the Haber-Bosch technique. In this catalytic process, N2 reacts with hydrogen under 1,200 degrees Celsius and 500 atm.

Since the production of chemical fertilizers requires large inputs of fossil fuel, chemical fertilizers can be relatively expensive.

The impact of the Haber-Bosch technology on agriculture has been very dramatic. The Haber-Bosch technology enables high-analysis ammonium fertilizers to be produced quickly. As a result, the reliance on biological N fixation and manures as N sources has declined.

Atmospheric nitrogen additions: Nitrogen is deposited onto the earth's surface by:

- Rain
  - o In the form of ammonium, nitrate, nitrite
- Finely divided organic N swept along the earth's surface
- Lightning
  - o Responsible for approximately 10-20 % of soil nitrate (114, Fertilizers)
- Industrial wastes
  - o In Hawaii, as compared to the major industrial regions of the Mainland, industrial wastes do not significantly contribute to atmospheric N.

Nitrogen Mineralization In Soils

When absorbed by plants, ammonium and nitrate are incorporated into plant cells as organic, or living, forms of nitrogen. When plants die, microorganisms break down, or decompose, dead plant cells. During the decomposition about plant matter, organic nitrogen is once again converted to inorganic ammonium and released into the soil. The process that converts organic N to ammonium is called mineralization and plays a significant role in the management of nitrogen.

Conditions affecting N mineralization: The amount of ammonium that is released to the soil through mineralization depends on several factors:

- Quantity of Organic Nitrogen: The amount of organic nitrogen originally present in the organic matter determines the amount of N that can ultimately be mineralized.
- Temperature: The optimal range for mineralization to occur is between 77-95 degrees Fahrenheit.
- Oxygen: Microorganisms need oxygen and since microorganisms mediate mineralization, sufficient oxygen must be available in the soil.
- Moisture content: Ideally, water should fill 15 - 70 % of pore space for maximum mineralization. This roughly corresponds to field capacity.
- Ratio of carbon to nitrogen (C:N): The C:N ratio is a term used to describe the relative amount of total carbon in comparison the amount of total nitrogen present in the soil and/or organic matter.
- This ratio is very important in determining the rate of mineralization that should occur for a given type of organic matter.

Since the microorganisms living in the soil need both carbon and nitrogen, net mineralization occurs when C:N ratio is less than 20:1. This means for every two parts of carbon, there should be 1 part nitrogen for net mineralization. If you are applying organic amendments to your soil, it is important to become familiar with the C:N ratio to ensure N availability.

After mineralization of N occurs, ammonium can be:

- Taken up by the plants
- Consumed by other organisms
- Nitrified
- Volatilized

Immobilization

Immobilization is process that converts inorganic nitrogen to organic nitrogen. It is the reverse reaction of mineralization.

Immobilization occurs when decomposing organic matter contains low amounts of nitrogen. Thus, immobilization occurs if the source of organic matter has a high C:N ratio. Microorganisms, who also need nitrogen to live, scavenge the soil for nitrogen when plant residues contain inadequate amounts of nitrogen. As inorganic ammonium and nitrate are incorporated into the cells of living microorganisms, the total N levels in the soil are reduced. Immobilization can ultimately result in nitrogen deficiencies.

When nitrogen is immobilized in the soil, there may be little nitrogen available for crop growth. As a result, plants can suffer from nitrogen deficiency and develop a yellow coloration. This is the reason why organic materials with a high C:N ratios, such as grass clippings and grain stover, are usually composted before they are incorporated into the soil or planting.

This allows time for soil microorganisms to decompose the materials and begin to release nitrogen and other nutrients back into the soil

Mineralization or Immobilization-

The processes of mineralization and immobilization are constantly occurring simultaneously. As organic matter decomposes, inorganic nitrogen will be released into the soil. As both plants and microorganisms grow, they utilize the nitrogen in the soil. Once plants and microorganisms die, they decompose and release inorganic nitrogen to the soil through mineralization.

Even though mineralization and immobilization are both occurring, we can determine which process, mineralization or immobilization, predominates.

- When mineralization occurs at a greater rate, we say that there is net mineralization.
- Likewise, when immobilization occurs to a greater extent, there is net immobilization.

What determines whether there is net mineralization or net immobilization-

The answer is the C:N ratio of the decomposable organic matter.

The C:N ratio is a characteristic of all organic matter, which includes:

- Crop residues
- Soil organic matter (including humus)
- Soil microorganisms (remember microorganism have both carbon and nitrogen)

Rule of thumb

- When the C:N ratio of decomposing organic residues is between 20:1 and 30:1, mineralization and immobilization occur at fairly equal rates.
- Net mineralization occurs at C:N ratios less than 20:1.
- Net immobilization occurs at C:N ratios greater than above 30:1.
- Most well decomposed organic matter in soils have a C:N ration near 10:1

Management of organic residues: If your program involves the addition of organic residues, it is important to know its C:N ratio. This knowledge allows you to predict whether net mineralization or net immobilization will occur. If the residue has a wide C:N ratio range, it may be necessary to apply additional amounts of nitrogen to your soil or choose a residue with a narrower range.

Nitrification

In most aerobic soils under optimal soil conditions, ammonium is rapidly converted to nitrate by soil bacteria through a process known as nitrification.

- Nitrification involves two steps:
  - o First, ammonium is converted to nitrite
  - o Then, nitrite is converted to nitrate.

As you can see from the outline of steps above, the intermediate product of nitrification is nitrite. If conditions are unfavourable to undergo the second step of nitrification, nitrite can leach into the ground water and pose as a health risk.

The process of nitrification produces hydrogen ions. When large quantities of ammonium-containing fertilizers are applied to soil over time, this process can acidify the soil. See figure below for a simplified presentation of the nitrification process.

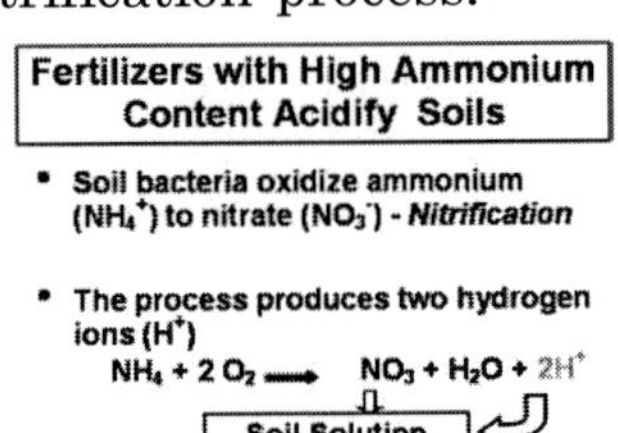

Figure: *Basic process that causes soil acidity by ammonium fertilizers.*

Source: Singleton, P. Nutrient Management Concepts: pH & Nutrient Formulation, University of Hawaii Cooperative Extension Service, Hilo Jul 25 2006.

Factors affecting nitrification: There are many factors that affect nitrification. Since nitrification s mediated by microorganisms, environmental factors that affect biological life will also influence nitrification. In general, the optimal conditions for most plant growth are also the optimal conditions for nitrification:

- Presence of ammonium in the soil: In order for nitrification to occur, there must be a source of ammonium in the soil. Sources include mineralized ammonium or additions of ammonium-containing synthetic fertilizers
- Presence of microorganisms: Microorganisms that carry out nitrification must be present in the soil.
- Soil pH: The optimal pH for nitrification is 8.5, but it may occur over a fairly wide pH range. However, acidity (less than 5.5) has a detrimental effect on the nitrifying bacteria, thus reducing nitrification.
- Soil moisture: Nitrification is optimal at the field capacity of the soil. Nitrification is reduced at moisture levels greater and below field capacity.
- Field capacity is the amount of water that remains in the soil after free drainage in a saturated soil ceases.
- Field capacity is also the optimal soil moisture for most plant growth.
- Soil aeration: Nitrification requires oxygen. Any management factor that improves soil aeration, such as adding organic matter, will help optimize nitrification.
- Soil temperature: Nitrifying bacteria are sensitive to temperature. The optimal temperature range for nitrification is between 77 and 95 degrees Fahrenheit. However, nitrification can occur between 41 and 95 degrees Fahrenheit.

Environmental Considerations: Nitrate is generally a very mobile in most soils. Excessive amounts of nitrate that are not taken up by plants is subject to leaching. Nitrate leaching can have an adverse effect on the environment.

- Nitrate, which moves through the soil profile during periods of rain, pollutes ground water reserves.

- Surface runoff of nitrate is a source of eutrophication, or algal blooms, in lakes and estuaries.

Gaseous Losses of Nitrogen

Denitrification

Denitrification is the biological process in which nitrate is converted to atmospheric N2.

- It is one source of N loss from the soil.
- Like other N processes, denitrification is a biological process that is mediated by denitrifying bacteria.

Soil conditions that lead to denitrification:

- Waterlogged soils: In waterlogged soils, the flow of air is poor. Even in aerated soils, small, localized areas in the soil (microsites) can lack oxygen. Any microenvironment within the soil that lacks oxygen is referred to as anaerobic. In contrast, nitrification requires oxygen and occurs under aerobic conditions.
- Presence of nitrate: Nitrate must be present for denitrification to occur. Nitrification provides a source of nitrate, as well as certain synthetic fertilizers.
- The greater the amount of nitrate present, the greater the denitrification potential.
- Presence of decomposable organic matter: Decomposing organic matter yields is a source of carbon. In return, carbon is the source of energy for denitrifying microorganisms.
- Oxygen: As stated earlier, denitrification occurs only when oxygen is absent. In aerated soils, denitrification can occur but is limited to those microsites that lack xygen.
- Soil pH: Denitrifying microorganisms are sensitive to low pH.
- Generally, denitrification is severely reduced at pH less than 5.0.
- The optimal range for denitrification is between 6.0 and 6.5.
- Soil Temperature: Denitrification will occur between 35 and 77 degrees Fahrenheit.

Volatilization: A second loss of nitrogen to the atmosphere is due to volatilization. By definition, volatilization is the loss of gaseous ammonia to the atmosphere.

Note the distinction between ammonia and ammonium. Although similar in form, ammonia is a gas that can escape from the soil into the atmosphere.

You may be familiar with ammonia, which is characterized by its pungent smell.

Factors Affecting Volatilization

There are several factors that affect volatilization:

- Soil pH: At a soil pH of 9.3, half of the ammonium in the soil is converted to ammonia and subject to volatilization loss. Generally, a pH greater than 7.5 allows for considerable loss of ammonia due to volatilization.
- Type of fertilizer: Urea fertilizers experience greater losses due to volatilization than ammonium fertilizers.
- However, if an ammonium fertilizer forms insoluble calcium compounds in the soil, the ammonium fertilizer will have greater volatilization losses than urea.
- Method of fertilizer placement: Broadcasting the fertilizer over the surface of the soil increases the losses due to volatilization. Incorporation into the soil reduces losses.
- Soil Temperature: The occurrence of volatilization increases as soil temperatures increase to 113 degrees Fahrenheit
- Soil Moisture: Evaporation promotes volatilization. Thus, volatilization is greatest as the soil dries after reaching field capacity.
- Buffering Capacity: Volatilization is less in well-buffered soils.
- Crop Residues: Crop residues that are not incorporated into the soil may increase the rate of volatilization.
- Manure: If not incorporated, nitrogen from manure sources can undergo volatilization.

Nitrogen Exchange And Nitrate Leaching

In our discussion on cation and anion exchange, we mentioned that cations, such as ammonium, are attracted to soil particles that have a cation exchange capacity. Since most surface soils have a cation exchange capacity, ammonium is retained by soil particles.

- Ammonium is largely immobile
- Losses of ammonium due to leaching are minimal

In contrast, nitrate is not retained by cation exchange capacity.

- Nitrate is highly immobile
- Losses of nitrate due to leaching can potentially be high

Not only is nitrate leaching an economic loss to the farmer, it is also an environmental concern. The conditions that lead to nitrate leaching follow:

- High rainfall intensity and distribution
- Highly irrigated fields
- Coarsely textured soils

Nitrate Leaching in Maui soils: In Hawaii, there are soils that can have a high anion exchange capacity under acidic conditions, which reduces nitrate leaching. This phenomenon primarily occurs in the acidic subsurface soil layers that have the greatest anion exchange capacity. As a result, researchers have found that subsurface soil layers in Hawaii can retain nitrate and may prevent it from leaching into the ground water.

Acidification: Management of Nitrogen Fertilizers: The fertilizer formulation can alter the pH of soil. Generally, fertilizers with high proportions of total nitrogen and are derived from ammonium sources (such as urea, ammonium sulfate, ammonium phosphate or ammonium nitrate) can acidify soils with repeated applications. Most fertilizers provide the "Lime Equivalent" on the bag's label. The lime equivalent is the amount of limestone (calcium carbonate) it takes to neutralize the acidifying effects of using one ton of a particular fertilizer.

In contrast, other fertilizers can increase soil pH. These fertilizers are usually low in ammonium, but high in nitrate. Additionally, these fertilizers sometimes contain calcium from calcium nitrate. The lime equivalent is also given for these fertilizers, but it indicates the equivalent liming effect rather than the lime needed to offset acidity.

The lime equivalent is only an estimate. The actual acidifying effect of the fertilizer is influenced, to some degree, by soil conditions that affect the transformation of the ammonium to nitrate and also by how much ammonium the plant assimilates before this transformation occurs. If growers know the history of their fertilizer applications over time, they can use the lime equivalent to predict when lime additions are likely needed.

See the table below for some fertilizer formulations and note the relationship between the percentage of nitrogen in the fertilizer that is derived from ammonium and the lime equivalent needed to counter this acidifying effect.

Table: Fertilizer composition can change the pH of soil

Fertilizer	Ammonium nitrogen	Fertilizer's Potential:	
		Acidity	Basicity
	%	*lbs lime/ton fertilizer*	
2-7-7	90	1700	
24-9-9	50	822	
20-20-20	69	583	
20–10-20	38	393	
20–0-20	25	40	
15-5-15	28		135

--data derived from various commercial fertilizer labels

Management Practices

Management practices that reduce N fertilization rates without affecting crop yields and quality are needed to reduce soil and environmental degradation, as soil degradation is directly related to increased N rates. Some of these practices include crop rotation, cover cropping, application of manure and compost, and integrated crop-livestock system. These practices can increase N inputs, reduce N fertilization rates, conserve soil organic matter, and enhance soil health and environmental quality without affecting crop yields compared with traditional management practices.

Crop rotation: Crop rotations that include legumes and nonlegumes in the rotation can substantially reduce N fertilization rates compared with nonlegume monocropping because legumes supply N to the soil due to their greater N concentration from atmospheric N fixation than nonlegumes. As no N fertilizer is applied to legumes, overall N fertilization rate is lower for the legume-nonlegume rotation than continuous nonlegumes while still maintaining crop yields. Sainju et al. observed that annualized crop biomass and grain yields under rainfed condition were similar or greater with legume-based rotations that included pea, durum (Triticum turgidum L.), canola (Brassica napus L.), and flax (Linum usitatissimum L.) than with continuous durum.

Crop rotation is an effective management practice to control weeds, diseases, and pests; reduce the risk of crop failure, farm inputs, and duration of fallow; and improve the economic and environmental sustainability of dryland cropping systems . Diversified crop rotations can efficiently use water and N compared with monocropping. For instance, wheat and barley can efficiently utilize soil water in wheat-pea and barley-pea rotations than continuous wheat and barley. This is because pea uses less water than wheat and barley, resulting in more water available for succeeding crops in the rotation.

Table 1. *Effect of crop rotation on average annualized crop biomass (stems and leaves) and grain yields of durum, canola, flax, and pea from 2006 to 2011 in eastern Montana, USA (Sainju et al., 2017d).*

Crop rotation†	Annualized biomass yield (Mg ha)	Annualized grain yield (Mg ha)
CD	3.32b‡	1.77a
D-C-D-P	4.02a	1.76a
D-D-C-P	3.90a	1.70a
D-F-D-P	3.39b	1.63ab
D-D-F-P	3.56b	1.54b

†Crop rotations are CD, continuous durum; D-C-D-P, durum-canola-durum-pea; D-D-C-P, durum-durum-canola-pea; D-F-D-P, durum-flax-durum-pea; and D-D-F-P, durum-durum-flax-pea.

‡Numbers followed by different letters within a column are significantly different at P 0.05 by the least square means test. Crop rotation can enhance or maintain soil organic C and N levels compared to monocropping. Both soil C and N stocks can be influenced by the quality and quantity of residue returned to the soil from crops involved in the rotation.

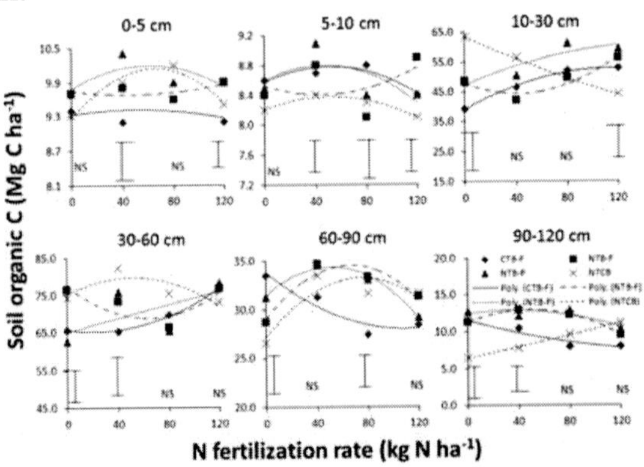

Figure: *Soil organic C at the 0-120 cm depth as affected by 6 years of N fertilization rates to malt barley in various cropping systems in eastern Montana, USA. CTB-F denotes conventional till malt barley-fallow; NTB-F, no-till malt barley-fallow; NTB-P, no-till malt barley-pea; and NTCB, no-till continuous malt barley. Vertical bars denote least significant difference between tillage and cropping sequence treatments within a N rate at P = 0.05 .*

Crop rotation can sequester C at 200 ± 120 kg C ha year, reaching equilibrium in 40-60 years compared with monocropping. Sainju found that soil organic C at 0-5 and 5-10 cm was similar in no-till malt barley-pea rotation (NTB-P) and no-till continuous malt barley (NTCB), both of which had greater soil organic C than no-till malt barley-fallow (NTB-F) and conventional till malt barley-fallow (CTB-F) due to greater amount of crop residue returned to the soil and reduced mineralization of soil organic matter. Similarly, Sainju et al. (2017d) found that soil total C at 0-125 cm was similar to continuous durum and rotations that included durum, canola, pea, and flax, except D-D-F-P. Soil total N at 0-120 cm was greater with spring wheat-pea rotation than continuous spring wheat.

Table 2. *Soil total C (STC) at the 0-125 cm depth after 6 years as affected by crop rotation in eastern Montana, USA.*

Crop rotation†	STC at 0–125 cm (Mg C ha)
CD	394.6a‡
D-C-D-P	395.4a
D-D-C-P	387.1a
D-F-D-P	395.4a
D-D-F-P	370.2b

†Crop rotations are CD, continuous durum; D-C-D-P, durum-canola-durum-pea; D-D-C-P, durum-durum-canola-pea; D-F-D-P, durum-flax-durum-pea; and D-D-F-P, durum-durum-flax-pea.

‡Numbers followed by different letters within a column are significantly different at P - 0.05 by the least square means test.

Table 3. *Soil total N (STN) at the 0-120 cm depth after 6 years as affected by crop rotation in eastern Montana, USA.*

Crop rotation	STN (Mg N ha)							
	0–5 cm	5–10 cm	10–20 cm	20–40 cm	40–60 cm	60–90 cm	90–120 cm	0–120 cm
CW	0.82	0.91	1.46	2.34b	2.11	2.29b	2.11	12.03b
W-P	0.85	0.90	1.53	2.66a	2.24	2.55a	2.23	12.96a
W-B-P	0.79	0.86	1.44	2.43ab	2.17	2..35b	2.22	12.17b
W-B-C-P	0.81	0.88	1.47	2.54a	2.26	2.51a	2.10	12.62ab

Crop rotations are CW, continuous spring wheat; W-P, spring wheat-pea; W-B-P, spring wheat-barley hay-pea; and W-B-C-P, spring wheat-barley hay-corn-pea.

Numbers followed by different letters within a column are significantly different at P - 0.05 by the least square means test.

In an experiment evaluating the effects of crop rotation and cultural practice (traditional and ecological) on N balance in dryland agroecosystems, Sainju et al. observed that N fertilization rates were lower with legume-based crop rotations (D-C-D-P, D-D-C-P, D-F-D-P, and D-D-F-P) than nonlegume monocropping (CD). Traditional cultural practices included conventional till, recommended seed rate, broadcast N fertilization, and reduced stubble height and ecological practices inlcuded no-till, increased seed rate, banded N fertilization, and increased stubble height.

They found that both total N input and output were greater with legume-based rotations than nonlegume monocropping due to pea N fixation and increased grain N removal. As a result, N balance was positive, indicating N surplus in legume-based rotations, and negative, indicating N deficit in nonlegume monocropping. This suggests that external N input is lower to sustain crop yields in legume-based crop rotations than nonlegume monocropping.

Table 4. *Annual N balance due to the difference between total N inputs and outputs and N sequestration rate under dryland agroecosystems from 2005 to 2011 in eastern Montana, USA [26, 27].*

Parameter	Traditional (kg N ha year)					Ecological (kg N ha year^{-1})				
	CD	D-C-D-P	D-D-C-P	D-F-D-P	D-D-F-P	CD	D-C-D-P	D-D-C-P	D-F-D-P	D-D-F-P
N inputs										
N fertilization rate	83A	62B	59B	52B	54B	87A	60B	63B	55B	56B
Pea N fixation	0C	84AB	76B	80AB	75B	0C	84AB	78B	87A	82AB
Atmospheric N deposition	14	14	14	14	14	14	14	14	14	14
N added by crop seed	3	3	3	3	3	3	3	3	3	3
Nonsymbiotic N fixation	5	5	5	5	5	5	5	5	5	5
Total N input	105B	167A	156A	154A	150A	109B	166A	162A	164A	159A
N outputs										
Grain N removal	49B	62A	57AB	54AB	55AB	52AB	65A	64A	63A	54AB
Denitrification	12	10	9	8	9	13	9	10	9	9
Ammonia volatilization	12	9	9	8	8	13	9	9	8	8
Plant senescence	5	7	6	6	6	6	7	7	7	6
N leaching	9	12	12	12	12	9	12	12	12	12
Gaseous N (NO_x) emissions	2	3	3	3	3	2	3	3	3	3
Surface runoff	1	2	1	1	1	1	2	2	2	2
Total N output	91B	105A	98AB	92B	94AB	96AB	107A	107A	103A	94AB
Changes in N level	14B	62A	58A	62A	56A	13B	59A	55A	61A	65A
N sequestration rate (0–125 cm)	50	45	42	46	43	52	48	46	44	40
N balance	−36 (±11)B	17 (±5)A	16 (±4)A	16 (±4)A	13 (±3)A	−39 (±12)B	11 (±3)A	9 (±2)A	17 (±4)A	25 (±5)A

a Crop rotation are CD, continuous durum; D-C-D-P, durum-canola-durum-pea; D-D-C-P, durum-durum-canola-pea; D-F-D-P, durum-flax-durum-pea; and D-D-F-P, durum-durum-flax-pea.

b Numbers followed by the same letter within a row are not significantly different at P - 0.05.

c Changes in N level = total N input - total N output.

d Determined from the linear regression analysis of soil total N (STN) at 0-125 cm from the year 2005 to 2011.

e N balance = changes in N levels - N sequestration rate (0-125 cm).

Legume-nonlegume rotation can also resist soil acidification compared with continuous nonlegumes. Sainju et al. reported that soil pH at 0-7.5 cm after 30 years of experiment initiation was 0.13-0.44 greater and at 7.5-15.0 cm was 0.11-0.29 greater with spring wheat-barley/pea rotation (FSTW-B/P) than continuous spring wheat (NTCW, STCW, and FSTCW).

They explained this as a result of lack of N fertilization to pea and reduced N fertilization rate to spring wheat following pea whose residue supplied N to spring wheat because of higher M concentration than spring wheat and barley residues. Soil residual NO3-N, which can pollute groundwater through leaching, was lower with legume-based crop rotations containing durum, canola, pea, and flax than continuous durum, suggesting that legume-based crop rotations can reduce N fertilization rate and the potential for N leaching compared with nonlegume monocropping.

Table 5. *Effect of tillage and crop rotation combination on soil pH and buffer pH at the 0-120 cm depth after 30 years of experiment initiation in eastern Montana, USA.*

Tillage and cropping sequence	**Soil depth**					
	0–7.5 cm	**7.5–15 cm**	**15–30 cm**	**30–60 cm**	**60–90 cm**	**90–120 cm**
pH						
NTCW	5.33ab	6.50abD	7.60C	8.35B	8.58A	8.75A
STCW	5.05bE	6.15bD	7.58C	8.25B	8.63A	8.70A
FSTCW	5.02bE	6.33bD	7.80C	8.30B	8.68AB	8.73A
FSTW-B/P	5.46aE	6.44bD	7.60C	8.15B	8.51A	8.59A
STW-F	5.73aE	7.03aD	7.65C	8.25B	8.50AB	8.66A
Contrast						
NT vs. T	0.29	0.26	−0.09	0.08	−0.08	0.04
CW vs. W-F	−0.68	−0.88	−0.08	0.01	0.13	0.04
CW vs. W-B/P	−0.43	−0.11	0.20	0.15	0.16	0.14
Buffer pH						
NTCW	6.45bE	7.10abD	7.43C	7.60B	7.70AB	7.73A
STCW	6.38bE	7.00bD	7.43C	7.58B	7.68A	7.70A
FSTCW	6.43bE	7.05bD	7.45C	7.60B	7.70AB	7.73A
FSTW-B/P	6.66aD	7.13abC	7.44B	7.58B	7.69AB	7.70A
STW-F	6.80aE	7.24aD	7.44C	7.59B	7.66AB	7.72A
Contrast						
NT vs. T	0.05	0.08	−0.01	0.01	0.01	0.01
CW vs. W-F	−0.43	−0.24	−0.01	−0.01	0.01	−0.01
CW vs. W-B/P	−0.24	−0.08	−0.01	0.03	0.01	0.03

a FSTCW, fall and spring till continuous spring wheat; FSTW-B/P, fall and spring till spring wheat-barley (1994-1999) followed by spring wheat-pea (2000-2013); NTCW, no-till continuous spring wheat; STCW, spring till continuous spring wheat; and STW-F, spring till spring wheat-fallow. CW represents continuous wheat; NT, no-till; T, till; W-B/P, spring wheat-barley/pea; and W-F, spring wheat-fallow.

b Numbers followed by the same lowercase letter within a column among treatments in a set are not significantly different at P - 0.05.

c Numbers followed by the same uppercase letter within a row among soil depths in a set are no significantly different at P - 0.05.

Table 6. *Soil NO3-N content at the 0-125 cm depth as affected by crop rotation and cultural practice averaged across years from 2006 to 2011 in eastern Montana, USA. Crop rotations are CD, continuous durum; DCDP, durum-canola-durum-pea; DDCP, durum-durum-canola-pea; DDFP, durum-durum-flax-pea; and DFDP, durum-flax-durum-pea. Numbers followed by different letters within a column are significantly different at P - 0.05 by the least square means test.*

Crop rotation	NO_3-N content at various depths (kg N ha)						
	0–5 cm	5–10 cm	10–20 cm	20–50 cm	50–88 cm	88–125 cm	0–125 cm
CD	2.47a	1.81a	2.43a	8.49a	9.37a	9.17a	33.87a
DCDP	1.82a	1.22b	1.94b	6.47a	7.77a	6.71b	26.32b
DDCP	1.86a	1.19b	1.93b	5.97a	8.07a	6.38b	25.59b
DFDP	1.90a	1.37b	2.20a	6.59a	9.62a	8.64ab	30.60a
DDFP	1.74a	1.28b	2.29a	6.27a	8.63a	6.65b	27.02b

Cover Cropping

Cover crops have been grown successfully in regions with mild winter to provide vegetative cover for reducing soil erosion. Cover crops are usually grown in the fall after the harvest of summer cash crops and have many benefits for sustaining crop yields and improving soil and water quality. Winter cover crops use soil residual N that may otherwise leach into groundwater after crop harvest in the fall, thereby reducing soil profile NO3-N content and N leaching. Summer cover crops are grown in the summer to replace fallow when no other crops are grown. Depending on the species, cover crops can maintain or increase soil organic C and N by providing additional crop residue which increases biomass C and N inputs to the soil and sequester atmospheric C and/or N, thereby reducing the rate of N fertilization to summer crops. Other benefits of cover crops include increased soil aggregation and water infiltration capacity , improved water holding capacity, and reduced soil erosion compared with no cover crop.

Integrating legumes in crop rotations can supply N to succeeding crops and increase crop yields compared to nonlegumes or no cover crop rotations. In contrast, nonlegume cover crops are effective in increasing soil organic C through increased biomass production compared with legumes or no cover crop. Nonlegumes also reduce NO3-N leaching from the soil profile better than legumes, or no cover crop do. As none of the cover crops are effective enough to provide most of these benefits, i.e., to supply N, sustain crop yields, increase soil organic matter, and reduce N leaching, a mixture of legume and nonlegume cover crops is ideal to supply both C and N inputs in adequate amounts that help to improve soil and water quality by increasing organic matter content and the potential for reducing N leaching compared with legumes and increase crop yields compared with nonlegumes.

Sainju et al. found higher biomass yield with hairy vetch/rye (Secale cereale L.) mixture than rye, hairy vetch, or winter weeds, and N concentration in the mixture similar to hairy vetch, except in 2001. As a result, they observed greater biomass C and N contents with hairy vetch/rye mixture than rye and winter weeds and similar to or greater than hairy vetch. The C/N ratio of cover crop biomass, which measures the decomposition rate of the residue, was similar between hairy vetch/rye mixture and hairy vetch.

Table 7. *Effect of cover crop species on aboveground biomass yield and C and N contents in cover crops from 2000 to 2002 in central Georgia, USA.*

Cover crop	**Biomass yield ($Mg\ ha^{-1}$)**	**Concentration**		**Content**		**C/N ratio**
		C ($g\ kg^{-1}$)	**N ($g\ kg^{-1}$)**	**C ($kg\ ha^{-1}$)**	**N ($kg\ ha^{-1}$)**	
2000						
Weeds	1.65d	370b	15b	587d	25d	24b
Rye	6.07b	430a	15b	2670b	68c	29a
Vetch	5.10c	394ab	33a	2006c	135b	12c
Vetch/rye	8.18a	366b	38a	3512a	310a	10c
2001						
Weeds	0.75d	391b	20b	277d	15b	20c
Rye	3.81b	448a	8d	1729b	32b	57a
Vetch	2.44c	398b	32a	964c	76a	12c
Vetch/rye	5.98a	434a	14c	2693a	84a	32b
2002						
Weeds	1.25c	375b	18b	476c	23b	21b
Rye	2.28b	434a	11b	986b	25b	40a
Vetch	5.16a	361b	36a	2094a	167a	10c
Vetch/rye	5.72a	381b	33a	2260a	186a	11c

†Cover crops are rye, cereal rye; vetch, hairy vetch; vetch/rye, hairy vetch and rye biculture; and weeds, winter weeds.

‡Numbers followed by the same letter within a column of a year are not significantly different at P - 0.05.

Because of increased C supply, soil organic C at 0-10 and 10-30 cm was also greater with hairy vetch/rye than other cover crops. At 30-60 cm, soil organic C was greater with hairy vetch/rye than other cover crops, except hairy vetch. Soil total N at 0-15, 15-30, and 0-120 cm was also greater with hairy vetch and hairy vetch/rye mixture than other cover crops. Similarly, soil residual NO3-N content at 0-120 cm was greater with hairy vetch than other cover crops and is slightly greater than that with 120-130 kg N ha-1 . Nitrogen loss at 0-120 cm during the winter fallow period from November to April was lower with hairy vetch/rye than other cover crops.

Nitrogen fertilizer equivalence of rye and winter weeds for cotton and sorghum ranged from -129 to 69 kg N ha-1, but those of hairy vetch and hairy vetch/rye ranged from 92 to 220 kg N ha-1 , suggesting that hairy vetch and hairy vetch/rye can increase cotton and sorghum yields similar to those by 92-220 kg N ha-1.

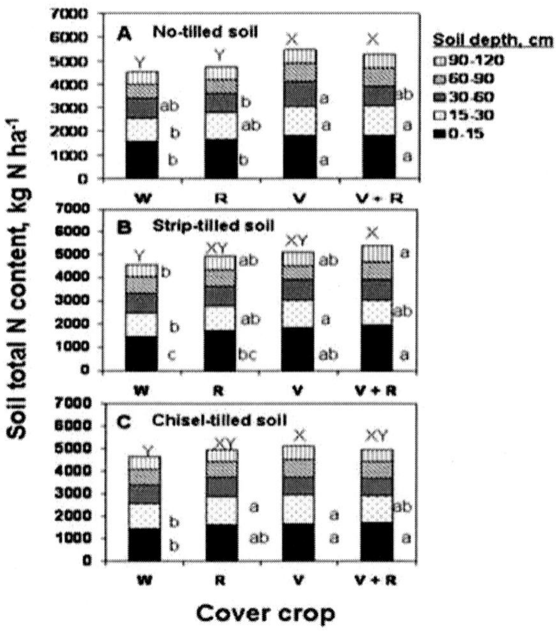

Figure: *Effect of cover crop on soil total N at the 0-120 cm depth in (A) no-tilled, (B) strip-tilled, and (C) chisel-tilled soils after 3 years in Central Georgia, USA. R denotes cereal rye; V, hairy vetch; V + R, hairy vetch and rye biculture; and WW, winter weeds. Bars followed by the same lowercase letter within a soil depth are not significantly different between cover crops at P = 0.05. Bars followed by the same uppercase letter at the top are not significantly different between cover crops at the 0-120 cm depth at P - 0.05 .*

These results suggest that hairy vetch/rye mixture can produce crop yields similar to hairy vetch. The mixture can also increase soil organic matter and reduce N fertilization rate and the potential for N leaching compared with rye and winter weeds. Therefore, legume-nonlegume cover crop mixture can provide several benefits, such as reducing the cost of N fertilization, maintaining crop yields, enhancing soil organic matter, and reducing N leaching compared with either cover crop alone or no cover crop.

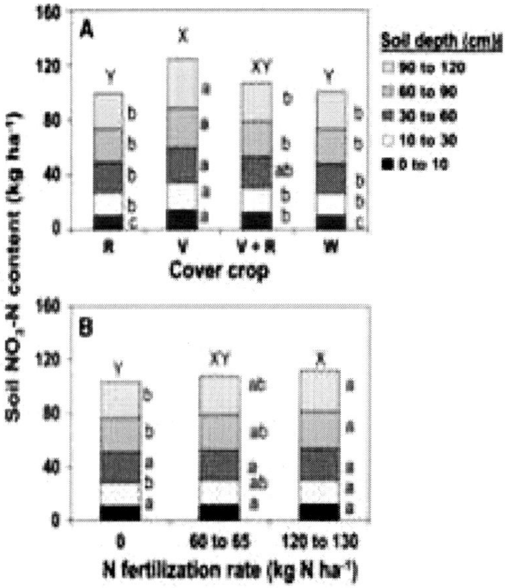

Figure: *Effect of (A) cover crop and (B) N fertilization rate on soil NO3-N content at the 0-120 cm depth in Central Georgia, USA. R, denotes cereal rye; V, hairy vetch; V + R, hairy vetch and rye biculture; and W, winter weeds. Bars followed by the same lowercase letter within a soil depth are not significantly different between cover crops at P = 0.05. Bars followed by the same uppercase letter at the top are not significantly different between cover crops at the 0-120 cm depth at P - 0.05.*

Table 8.

Cover crop	Total crop residue and soil N (kg N ha^{-1})			Total crop residue and soil N (kg N ha^{-1})		
	November 2000	**April 2001**	**Loss**	**November 2001**	**April 2002**	**Loss**
Rye	5057bc	4888b	169b	4820b	4764b	56a
Vetch	5455a	5235a	220a	5323a	5244a	79a
Vetch/rye	5249ab	5141a	108c	5222a	5182a	40a
Weeds	4869c	4709b	160b	4725b	4649b	76a

Effect of cover crop on N loss from crop residue and soil N (NH4-N + NO3-N + organic N contents) at the 0-120 cm depth during the two winter seasons (from November 2000 to April 2001 and from November 2001 to April 2002) in central Georgia, USA.

†Cover crops are rye, cereal rye; vetch, hairy vetch; vetch/rye, hairy vetch and rye biculture; and weeds, winter weeds or no cover crop.

‡Include soil NH4-N + NO3-N + organic N contents at 0-120 cm, and N returned to the soil from cotton biomass (stems + leaves) in November 2000 and cover crop biomass in April 2001.

§Include soil NH4-N + NO3-N + organic N contents at 0-120 cm, and N returned to the soil from sorghum biomass (stems + leaves) in November 2001 and cover crop biomass in April 2002.

Numbers followed by the same letter within a column are not significantly different at P - 0.05.

Table 9.

Parameter	Cover crop				Regression analysisa	
	Winter weeds	Rye	Hairy vetch	Hairy vetch/rye	R^2	*P*
2000 cotton						
Lint yield	–	–	–	–	0.25	0.67
Lint N uptake	–	–	–	–	0.25	0.67
Biomass yield	−13	30	149	93	0.96	0.13
Biomass N uptake	−21	2	165	92	0.99	0.06
Soil inorganic N	−60	−190	220	140	0.64	0.40
2001 sorghum						
Lint yield	7	−64	107	179	0.96	0.12
Lint N uptake	25	−67	167	150	0.96	0.14
Biomass yield	32	−168	194	194	0.99	0.02
Biomass N uptake	69	−84	192	83	0.98	0.08
Soil inorganic N	59	12	116	71	0.86	0.25
2002 cotton						
Lint yield	–	–	–	–	0.28	0.82
Lint N uptake	–	–	–	–	0.24	0.87
Biomass yield	−21	−61	139	205	0.96	0.12
Biomass N uptake	−35	−13	134	160	0.97	0.11
Soil inorganic N	−74	5	176	160	0.70	0.37

Nitrogen fertilizer equivalence (kg N ha-1) of cover crops and soil inorganic N (NH4-N + NO3-N) content at the 0-30 cm depth for cotton and sorghum yields and N uptake from 2000 to 2002 in central Georgia, USA. aRegression analysis of N fertilization rates versus cotton and sorghum yields and N uptake and soil inorganic N.

Application of manure and compost: Manure and compost are rich sources of nutrients, and their application can increase soil organic C and total N, improving soil quality and crop production compared to no fertilizer application. Sainju et al. compared soil organic C and total N after 10 years of poultry litter with inorganic N fertilizer applications, both applied at 100 kg N ha-1 to corn and cotton). They found that soil organic C and total N at 0-20 cm were greater with poultry litter application than inorganic N fertilization, regardless of tillage practices. As a result, poultry litter application sequestered C at 461 kg C ha-1 year-1 and N at 38 kg N ha-1 year-1 compared to 38 kg C ha-1 year-1 and 4 kg N ha-1 year-1, respectively, with N fertilization. As poultry litter also supplied C at 1.7 Mg C ha-1 year-1 and only 60% of N from poultry litter was available to crops in the first year, Sainju et al. reported that part of non-mineralized C and N from the litter converted to soil organic C and N, thereby increasing their levels with poultry litter application. In contrast, little or no C was supplied by inorganic N fertilizer, and most of N supplied by the fertilizer can either be taken up by the crop or lost to the environment through leaching, denitrification, and volatilization.

Table 10. *Effect of tillage and N source on soil organic C (SOC) at the 0-20 cm depth after 10 years in Alabama, USA.*

Tillage	N source	SOC concentration (g C kg⁻¹)		SOC content (Mg C ha⁻¹)	Changes in SOC from 1996 to 2006 (Mg C ha⁻¹)	C sequestration rate (kg C ha⁻¹ year.⁻¹)
	100 kg N ha⁻¹	0–10 cm	10–20 cm	0–20 cm	0–20 cm	0–20 cm
NT	AN	13.5	11.0	40.1	1.47	147
	PL	15.9	10.5	43.7	5.10	510
MT	AN	15.9	11.0	42.6	3.97	397
	PL	15.4	10.6	42.2	3.63	363
CT	AN	14.3	10.7	37.4	- 1.20	- 120
	PL	15.3	11.8	43.7	5.10	510
LSD (0.05)		—	—	3.1	3.1	310
Means	AN	14.6a	10.9a	40.0b	1.41b	141b
	PL	15.6a	11.0a	43.2a	4.61a	461a

†Tillage is CT, conventional till; MT, mulch till; and NT, no-till.

‡N source is AN, NH4NO3; and PL, poultry litter. § Numbers followed by different letters within a column in a set are significantly different at P - 0.05 by the least square means test.

Table 11. *Effects of tillage and N source on soil total N and N sequestration rate at the 0-20 cm depth after 10 years in Alabama, USA.*

Tillage	N source	STN concentration (g N kg⁻¹)		STN content (Mg N ha⁻¹)	Change in STN from 1996 to 2006 (Mg N ha⁻¹)	N sequestration rate (kg N ha⁻¹ year⁻¹)
	(100 kg N ha⁻¹)	0–10 cm	10–20 cm	0–20 cm	0–20 cm	0–20 cm
NT	AN	1.23	1.03	3.44	−0.23	−23
	PL	1.52	1.02	4.19	0.49	49
MT	AN	1.42	1.01	3.84	0.15	15
	PL	1.49	0.92	3.91	0.21	21
CT	AN	1.31	0.98	3.67	−0.03	−3
	PL	1.51	1.04	4.11	0.41	41
LSD (0.05		—	—	0.24	0.24	24
Means	AN	1.55b	1.59a	3.65b	−0.04b	−4b
	PL	1.65a	1.59a	4.07a	0.38a	38a

aTillage is CT, conventional till; MT, mulch till; and NT, no-till.

bN source is AN, ammonium nitrate; and PL, poultry litter.

cLeast significant differences between treatments at P = 0.05.

dNumbers followed by the same letter within a column in a set are not significantly different at P - 0.05.

Because of lower N availability from poultry litter as a result of reduced N mineralization, total aboveground biomass and N uptake of corn, cotton, and rye cover crop were lower with poultry litter application than inorganic N fertilization. Although soil health and quality can be improved with poultry litter application through organic matter enrichment, crop yields can be lower compared with N fertilization.

For enhancing soil and environmental quality and sustaining crop yields, both inorganic N fertilizer and manure/compost should be applied as a mixture in balanced proportion as per crop demand after analyzing soil NO3-N test to a depth of 60 cm. This could reduce N fertilization rate and undesirable consequences of N fertilization on soil and environmental quality.

Table 12. *Effects of cropping system and N source on total biomass (stems + leaves) residues of rye, cotton, and corn and N uptake from 1997 to 2005 in Alabama, USA. †Numbers followed by the same letter within a column in a set are not significantly different at P - 0.05.*

Cropping system	N source	Total crop biomass	Total N uptake
	100 kg N ha^{-1}	(Mg ha^{-1})	(kg N ha^{-1})
Rye/cotton-rye/cotton-corn		137.0a	1544a
Cotton-cotton-corn		110.2b	1247b
	NH_4NO_3	133.3a	1502a
	Poultry litter	111.8b	1289b

Integrated crop-livestock system: Integrated crop-livestock systems were commonly used to sustain crop and livestock products throughout the world before commercial fertilizers were introduced in 1950. The system is still common among producers in developing countries, especially in Africa and Asia where fertilizers are scarce and expensive. The integrated crop-livestock system has the potential to improve soil quality and sustain crop yields. The major benefits of the system are (1) production of crops, meat, and milk, (2) production of crop residue for animal feed, (3) production of manure to apply as fertilizer, (4) use of animals as draft power for tillage, and (5) control of weeds and pests.

Animal grazing during fallow periods in wheat-fallow systems can be used to effectively control weeds and insects, such as wheat stem saw fly [Cephus cinctus Norton (Hymenoptera: Cephidae)]. The animal usually grazes on crop residues and weeds during the fallow period. Although grazing can reduce the quantity of crop residue returned to the soil, the number of animals grazed per unit area can be adjusted in such a way that crop residue cover in the grazing treatment will be similar to that in the conservation tillage system where soil erosion is minimal . Animal feces and urine returned to the soil during grazing can enrich soil nutrients, improve soil quality, and increase crop yields . The distribution of feces and urine by animals during grazing at the soil surface can be uneven; however, distribution can be more uniform with sheep than with cattle grazing.

Hatfield et al. reported that sheep grazing during fallow did not affect soil organic matter and nutrient levels compared to the non-grazed treatment in the North Central Montana. Sheep grazing can increase soil bulk density and extractable P and grass yields compared to cattle grazing. Snyder et al. found similar or greater wheat grain yields with and without animal grazing. Similarly, Quiroga et al. observed that 10 years of cattle grazing did not alter soil P concentration in Argentina. In contrast, Niu et al. in Australia observed greater soil P and K concentrations in sheep camping than in non-camping sites due to increased animal excreta. Cattle and sheep grazing in the pasture can increase soil P and K concentrations compared to non-grazing.

Sainju et al.reported that annualized wheat grain and biomass yields were lower with spring wheat-fallow and winter wheat-fallow rotations than continuous spring wheat due to the absence of crops during the fallow period (Table 13). In contrast, wheat grain yield was not different among weed management practices where sheep grazing was used among one of the treatments to control weeds along with herbicide application and tillage, although wheat biomass yield was lower with sheep grazing and herbicide application than tillage. Soil organic C,total N, and NO3-N contents varied among weed management practices and soil depths, but the contents at 0-120 cm were not affected by weed management practices. Soil P, K, and SO4-S contents at 0-30 cm were lower with sheep grazing than other weed management practices, but pH, electrical conductivity, and Ca, Mg, and Na contents were similar or greater with sheep grazing. Consumption of crop residue by sheep during grazing, but little P and K inputs to the soil through urine and feces, reduced soil P and K concentrations with sheep grazing compared with other weed management practices. These results suggest that sheep grazing can reduce the cost of animal feed without seriously affecting crop yields and sustain soil organic matter and nutrients compared with other weed management practices, except P and K which need to be added with inorganic fertilizers to eliminate their deficiency.

As soil residual NO3-N content was not different among weed management practices, long-term study may be needed to evaluate if animal grazing can reduce N fertilization rate for crop production. However, animal grazing can recycle nutrients and control weeds effectively compared with herbicide application and tillage, thereby saving the cost of fertilization and weed control.

Year	Cropping sequence (Mg ha^{-1})			Weed management (Mg ha^{-1})			Mean
	CSW	SW-F	WW-F	Chem.	Mech.	Graz.	
Annualized grain yield							
2004	5.55aA	2.90aC	3.53aB	3.92aA	4.01aA	4.05aA	3.99a
2005	2.68bA	1.83bB	1.15eC	1.84cA	1.92bA	1.90bA	1.89b
2006	2.57bA	1.45cB	1.70 dB	1.89cA	1.90bA	1.92bA	1.90b
2007	1.86cB	1.18cC	2.95bA	1.89cA	2.03bA	2.00bA	2.00b
2008	2.61bA	1.56bcC	2.22cB	2.09bA	2.17bA	2.14bA	2.13b
Mean	3.05A	1.78C	2.31B	2.32A	2.42A	2.40A	

Legumes in the crop rotation can supply N from its residue to succeeding crops, thereby reducing N fertilization rates to succeeding nonlegumes. Also diversified crop rotations can use N and water more efficiently and reduce weed, pest, and disease infestations, thereby enhancing crop yields compared with continuous nonlegume monocropping. Cover crops grown to replace the fallow period can reduce soil erosion, enhance soil organic matter, and help to enrich soil health and fertility. Legume covers crop supply N and reduce N fertilization rate. Application of manure and compost can also enhance soil health and quality; however, additional inorganic N fertilization at lower rate is required to sustain crop yield and quality.

Similarly, integrated crop-livestock system can help to reduce N fertilization rate by returning N and other nutrients through urine and feces to the soil during animal grazing without affecting crop yields. Some additional N fertilizer, however, may be required for sustainable crop production, because animals return only a part of nutrients through urine and feces to the soil, while most of the crop residue grazed is used to increase the live weight of the animal. The choice of the management practice to reduce N fertilization rate to crops depends on soil and climatic conditions and social, cultural, and economic perspectives of the producers.

Liming: Soil acidification can be reduced by applying lime. However, lime is bulky and requires in large amount to neutralize soil acidity. The transportation cost to carry lime from manufactures to farms is also high and especially so in hilly regions where roads are few or lacking. As a result, it is expensive to apply lime and most producers in developing countries cannot afford to apply it. Furthermore, neutralization of soil acidity with lime application is only temporary in nature. This suggests that lime should be applied frequently to neutralize acidity, which increases the cost of production. The best practice to reduce soil acidity is to reduce the rate of N fertilization. Several management practices, such as legume-nonlegume crop rotation, cover cropping, application of manures and compost, and integrated crop-livestock system, can reduce N fertilization rate without affecting crop yields.

Degradation in soil and environmental quality can be mitigated, and crop yields can be sustained by reducing N fertilization rates and using novel management techniques that increase N cycling and N-use efficiency. These techniques include legume-nonlegume crop rotation, cover cropping, application of manures and compost, and integrated crop-livestock system. Soil acidity can be neutralized by lime application, but the effect is temporary. It is expensive to apply lime, and many producers in developing countries cannot afford to do so. Adaptation of these techniques to specific places depends on soil and climatic conditions and social, cultural, and economic perspectives of the producers.

Chapter 4

Phosphorus: Role, Dynamics and Management

Introduction to Phosphorus

Phosphorus is an essential nutrient, both as a part of several key plant structure compounds and as a catalysis in the conversion of numerous key biochemical reactions in plants. Phosphorus is noted especially for its role in capturing and converting the sun's energy into useful plant compounds.

Phosphorus is a vital component of:

- DNA, the genetic "memory unit" of all living things. It is also a component of RNA, the compound that reads the DNA genetic code to build proteins and other compounds essential for plant structure, seed yield and genetic transfer. The structures of both DNA and RNA are linked together by phosphorus bonds.
- ATP, the "energy unit" of plants. ATP forms during photosynthesis, has phosphorus in its structure, and processes from the beginning of seedling growth through to the formation of grain and maturity.

Thus, phosphorus is essential for the general health and vigor of all plants. Some specific growth factors associated with phosphorus are:

- Stimulated root development
- Increased stalk and stem strength
- Improved flower formation and seed production
- More uniform and earlier crop maturity

- Increased nitrogen N-fixing capacity of legumes
- Improvements in crop quality
- Increased resistance to plant diseases
- Supports development throughout entire lifecycle

Phosphorus Deficiency in Plants

Phosphorus deficiency is more difficult to diagnose than that of nitrogen or potassium.

Crops usually display no obvious symptoms of phosphorus deficiency other than a general stunting of the plant during early growth. By the time a visual deficiency is recognized, it may be too late to correct in annual crops.

Some crops, such as corn, tend to show an abnormal discolouration when phosphorus is deficient. The plants are usually dark bluish-green in colour with leaves and stem becoming purplish.

The degree of purple is influenced by the genetic makeup of the plant, with some hybrids showing much greater discolouration than others. The purplish colour is due to accumulation of sugars that favours the synthesis of anthocyanin (a purplish-coloured pigment), which occurs in the leaves of the plant.

Phosphorus is highly mobile in plants and, when deficient, it may be translocated from old plant tissue to young, actively growing areas. Consequently, early vegetative responses to phosphorus are often observed. As a plant matures, phosphorus is translocated into the fruiting areas of the plant, where high-energy requirements are needed for the formation of seeds and fruit.

Phosphorus deficiencies late in the growing season affect both seed development and normal crop maturity. The percentage of the total amount of each nutrient taken up is higher for phosphorus late in the growing season than for either nitrogen or potassium.

The level of moisture in the soil at the time of application has a significant effect on how much P is available to the plant during the season.

Symptoms in Corn

The photo displays a P deficient corn plant. Older leaves are affected before younger ones because of the redistribution of P in the plant. Corn may display a purple or reddish colour on the lower leaves and stems.

This condition is associated with accumulation of sugars in P-deficient plants, especially during times of low temperature.

Figure: *Phosphorus In Soils*

The total phosphorus content of most surface soils is low, averaging only 0.6% phosphorus. This compares to an average soil content of 0.14% nitrogen and 0.83% potassium.

The phosphorus content of soils is quite variable, ranging from less than 0.04% P-O- in the sandy soils of the Atlantic and Gulf coastal plains to more than 0.3% in soils of the northwestern United States.

Many factors influence the content of soil phosphorus:

- Type of parent material from which the soil is derived
- Degree of weathering and erosion
- Climatic conditions
- Crop removal and fertilization

Organic Phosphorus: Organic phosphorus is found in plant residues, manures and microbial tissues. Soils low in organic matter may contain only 3% of their total phosphorus in the organic form, but high-organic-matter soils may contain 50% or more of their total phosphorus content in the organic form.

Inorganic Phosphorus: Inorganic forms of soil phosphorus consist of apatite (the original source of all phosphorus), complexes of iron and aluminum phosphates, and phosphorus absorbed onto clay particles.

The solubility of these phosphorus compounds as well as organic phosphorus is extremely low, and only very small amounts of soil phosphorus are in solution at any one time. Most soils contain less than 1 pound per acre of soluble phosphorus, with some soils containing considerably less.

Through adequate phosphorus fertilization and good crop/soil management, soil solution phosphorus can be replaced rapidly enough for optimum crop production.

Soil Phosphorus Availability: Soluble phosphorus, either from fertilizer or natural weathering, reacts with clay, iron and aluminum compounds in the soil, and is converted readily to less available forms by the process of phosphorus fixation.

Because of these fixation processes:

- Phosphorus moves very little in most soils (less than an inch)
- Phosphorus stays close to its place of origin
- Cops seldom absorb more than 20 percent of fertilizer phosphorus during the first cropping season after application

As a result, little soil phosphorus is lost by leaching. This fixed, residual phosphorus remains in the rooting zone and will be slowly available to succeeding crops. Soil erosion and crop removal are the significant ways soil phosphorus is lost.

Factors of Phosphorus Availability

Soil Ph

Precipitation of phosphorus as slightly soluble calcium phosphates occurs in calcareous soils with pH values around 8.0. Under acid conditions, phosphorus is precipitated as Fe or Al phosphates of low solubility. Maximum availability of phosphorus generally occurs in a pH range of 6.0 to 7.0.

This is one of the beneficial effects of liming acid soils. Maintaining a soil pH in this range also favours the presence of H-PO-- ions, which are more readily absorbed by the plant than HPO-- ions, which occur at pH values above 7.0.

Balanced Crop Nutrition

Adequate supplies of other plant nutrients tend to increase the absorption of phosphorus from the soil:

- Application of ammonium forms of nitrogen with phosphorus increases phosphorus uptake from a fertilizer as compared to applying the phosphorus fertilizer alone or applying the nitrogen and phosphorus fertilizers separately.
- Applications of sulfur often increase the availability of soil phosphorus on neutral or basic soils, where the soil phosphorus is present as calcium phosphates.

Organic Matter: Soils high in organic matter contain considerable amounts of organic phosphorus that are mineralized (similar to organic nitrogen), and provide available phosphorus for plant growth. In addition to supplying phosphorus, organic matter also acts as a chelating agent and combines with iron, thereby preventing the formation of insoluble iron phosphates.

Heavy applications of organic materials such as manure, plant residues or green manure crops to soils with high pH values not only supply phosphorus, but upon decomposition, provide acidic compounds, which increase the availability of mineral forms of phosphorus in the soil.

Type Of Clay: Clay particles tend to retain or fix phosphorus in soils. Consequently, fine-textured soils such as clay loam soils have a greater phosphorus-fixing capacity than sandy, coarse-textured soils. Clays of the 1:1-type (kaolinite) have a greater phosphorus-fixing capacity than the 2:1-type clays (montmorillonite, illite, vermiculite).

Soils formed under high rainfall and high temperatures contain large amounts of kaolinitic clays, and therefore have a much greater fixing capacity for phosphorus than soils containing the 2:1-type clay. High temperatures and high rainfall also increase the amount of iron and aluminum oxides in the soil, which contributes greatly to the fixation of phosphorus added to these soils.

Application Timing: Fixation of soil phosphorus increases with time of contact between soluble phosphorus and soil particles. Consequently, more efficient utilization of fertilizer phosphorus is generally obtained by applying the fertilizer shortly before planting the crop. This practice is especially effective on soils with high phosphorus-fixing capacities.

On coastal plain areas, fertilizers may be applied several months before planting with little or no decrease in availability of the fertilizer phosphorus to the crop. Banding of fertilizer for row crops is also much more likely to increase the efficiency of fertilizer phosphorus on soils of high phosphorus-fixing capacity than on soils of low phosphorus-fixing capacity.

Soil Temperature, Aeration, Moisture And Compaction

Phosphorus absorption by the plant is decreased by low soil temperature and poor soil aeration. Starter fertilizers containing water-soluble phosphorus are much more likely to increase crop growth during cool weather. Excessive soil moisture or soil compaction reduces the soil oxygen supply and decreases the ability of the plant roots to absorb soil phosphorus.

Compaction reduces aeration and pore space in the root zone. This reduces phosphorus uptake and plant growth. Compaction also decreases the soil volume that plant roots penetrate, limiting their total access to soil phosphorus.

Soil Test Phosphorus Levels: Crop responses to fertilizer phosphorus will be greater and occur more frequently on soils testing low in phosphorus than on high-testing soils. However, yields on soils with high P soil test levels usually are higher. The response to phosphorus fertilizer on high-testing soils is increasing, and it is important to maintain high soil phosphorus levels to support optimum crop production.

Phosphorus Placement

If a grower is looking for maximum return from high phosphorus investment on low-testing soils, band application is best. Where conservation tillage is practiced, combinations of band and broadcast applications of phosphorus may be needed. This ensures an early, accessible phosphorus supply for developing seedlings and a nutrient reserve later in the growing season, when phosphorus demands remain strong.

Advantages Of Broadcast And Plow-down Phosphorus Applications

- High rates can be applied without injuring the plant
- Nutrient distribution throughout the root zone encourages deeper rooting, while band placement causes root concentration around the band
- Deeper rooting permits more root-soil contact, providing a larger reservoir of moisture and nutrients
- Practical way to apply fertilizer to forages
- Helps ensure full-feed fertility to help the crop take full advantage of favourable growth conditions throughout the growing season

Dual application of anhydrous ammonia and ammonium polyphosphates at seeding of wheat has been found to be superior to broadcast or band applications of ammonium polyphosphates.

Placement directly under the drill row (band seeding) for forage crops has proven superior to broadcast or side placement.

Adapted from "The Efficient Fertilizer Use Manual," Phosphorous chapter by Dr. Bill Griffith

Phosphorus Forms and Functions

- The orthophosphates, H2PO4- and HPO42-, are the primary forms of phosphorus taken up by plants.
- When the soil pH is less than 7.0, H2PO4- is the predominate form in the soil.
- Although less common, certain organic phosphorus forms can also be directly taken up by plants.

Functions of Phosphorus in Plants

Phosphorus is involved in many plant processes, including:

- Energy transfer reactions
- Development of reproductive structures
- Crop maturity
- Root growth
- Protein synthesis

The Phosphorus Cycle: In contrast to nitrogen, the atmosphere does not provide phosphorus. Instead, orthophosphates originate largely from primary and secondary minerals and/or from organic sources. However, the phosphorus cycle is by no means less complex than the nitrogen cycle, and there are many factors that affect the availability of phosphorus in the soil. The diagram below is an illustration of the phosphorus cycle.

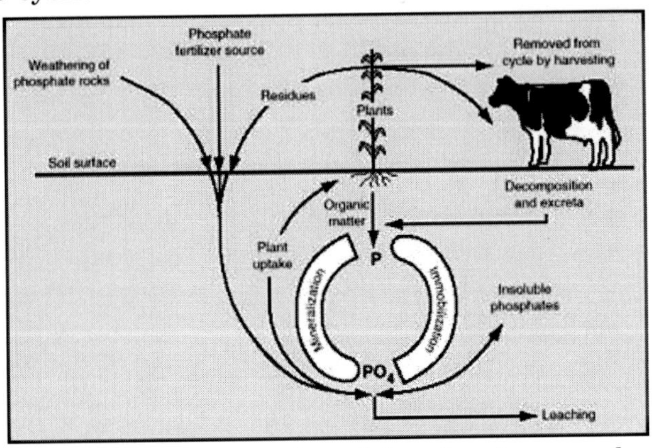

Figure: *A representation of the phosphorus cycle.*

Phosphorus Uptake By Plant Roots: Plant roots absorb phosphorus from the soil solution. In comparison to other macronutrients, the phosphorus concentration in the soil solution is much lower and ranges from 0.001 mg/L to 1 mg/L (Brady and Weil, 2002). In general, roots absorb phosphorus in the form of orthophosphate, but can also absorb certain forms of organic phosphorus. Phosphorus moves to the root surface through diffusion.

However, the presence of mycorrhizal fungi, which develop a symbiotic relationship with plant roots and extend threadlike hyphae into the soil, can enhance the uptake of phosphorus, as well especially in acidic soils that are low in phosphorus.

Phosphorus Sorption And Desorption: P-sorption occurs when the orthophosphates, H2PO4- and HPO42-, bind tightly to soil particles.

Since phosphate is an anion, particles that generate an anion exchange capacity will form strong bonds with phosphate.

Particles with anion exchange capacity:

- Aluminum and iron oxides
- Highly weathered kaolin clays (under acidic conditions)
- Amorphous materials.

These particles are commonly found in many of the most highly weathered soils and high weathered volcanic soils of Hawaii. Since P-sorption results in a decrease of plant available phosphorus, P-sorption can become a major issue in many Hawaii soils.

Additionally, in calcareous soils P-sorption may occur as phosphates sorb to impurities such as aluminum and iron hydroxides or displace carbonates in calcium carbonate minerals.

Factors that Affect P-Sorption

- Soil Mineral Type: Mineralogy of the soil has a great effect on P-sorption.
    - o Volcanic soils tend to have the greatest P-sorption of all soils since volcanic soils contain large amounts of amorphous material.
    - o Following volcanic soils, highly weathered soils (such as Oxisols and Ultisols) have the next greatest P-sorption capacities. This is due to the presence of large amounts of aluminum and iron oxides and highly weathered kaolin clays.

  - On the other end of the spectrum, less weathered soils and organic soils have low P-sorption capacities.
- Amount of clay: As the amount of clay increases in the soil, the P-sorption capacity increases as well. This is because clay particles have a tremendous amount of surface area for which phosphate sorption can take place.
- pH: At low pH, soils have greater amounts of aluminum in the soil solution, which forms very strong bonds with phosphate. In fact, a soil binds twice the amount of phosphorus under acidic conditions, and these bonds are five times stronger.
- Temperature: Generally, P-sorption increases as temperature increases.

Factors that decrease P-sorption:

- Other anions, such as silicates, carbonates, sulfates, arsenate, and molybdate, compete with phosphate for a position on the anion exchange site. As a result, these anions can cause the displacement, or desorption, of phosphate from the soil exchange site. Desorption causes phosphate availability in the soil solution to increase.
- Organic matter increases P availability in four ways.
  - First, organic matter forms complexes with organic phosphate which increases phosphate uptake by plants.
  - Second, organic anions can also displace sorbed phosphate.
  - Third, humus coats aluminum and iron oxides, which reduces P sorption.
  - Finally, organic matter is also a source of phosphorus through mineralization reactions.
- Flooding the soil reduces P-sorption by increasing the solubility of phosphates that are bound to aluminum and iron oxides and amorphous minerals.

Phosphate Precipitation and Dissolution

Phosphate precipitation is a process in which phosphorus reacts with another substance to form a solid mineral.

In contrast, dissolution of phosphate minerals occurs when the mineral dissolves and releases phosphorus.

Precipitation and dissolution reactions greatly influence the availability of phosphate in the soil.

- Phosphate minerals can dissolve over time to replenish the phosphate in the soil solution. This reaction increases the availability of phosphorus.
- On the other hand, phosphate minerals form by removing phosphate from soil solution. This reaction decreases the availability of phosphorus.
- However, both precipitation and dissolution are very slow processes.

Solubility of Phosphate Minerals: The solubility of phosphate minerals is very dependent upon soil pH.

- The soil pH for optimum phosphorus availability is 6.5
- At high or neutral pH, phosphate reacts with calcium to form minerals, such as apatite.
- Under acidic conditions, phosphorus may react with aluminum and iron to form minerals, such as strengite and varescite.

Mineralization and Immobilization of Phosphate

In an average soil, approximately 50% of total phosphorus is organic. Thus, soil organic phosphorus is a very important aspect of the P cycle.

The various sources of organic phosphorus include:

- Phytin
- Nucleic acids
- Phospholipids

Like nitrogen, organic phosphorus is converted to inorganic phosphate through the process of mineralization.

The immobilization of inorganic phosphate, in contrast, is the reverse reaction of mineralization. During immobilization, microorganisms convert inorganic forms to organic phosphate, which are then incorporated into their living cells.

Mineralization and immobilization of phosphorus occur simultaneously in the soil. Ultimately, the C:P ratio determines whether there is net mineralization or net immobilization.

- When the C:P ratio is less than 200:1, net mineralization prevails. Net mineralization indicates that there is enough phosphorus in the soil to sustain both plants and microorganisms.

- When the C:P ration is between 200:1 and 300:1, immobilization and mineralization rates are fairly equal.
- When the C:P ratio is greater than 300:1, net immobilization occurs. During immobilization there is not enough P to sustain both plants and microorganisms; and so, microorganisms scavenge the soil for P.

Factors affecting mineralization and immobilization

The factors that affect P mineralization and immobilization are the same that affect nitrogen mineralization and immobilization:

- Temperature
- Moisture
- Aeration

Management of phosphorus-P-fixation: P-fixation is a term that is used to describe both P-sorption and P precipitation. Since both P-sorption and P precipitation reduce phosphorus availability, a soil with a great P-fixation capacity has less available phosphorus after fertilization than a soil with a low P-fixation capacity.

In other words, when the same amount of fertilizer is applied to a volcanic soil and a moderately weathered grassland soil, the volcanic soil has less P available due to its greater P-fixation capacity.

Soil Phosphorus Management Options for Sustainable Crop Production

Sustainable crop production aims at maintaining high crop yield without adversely affecting ecosystems to meet the need of current as well as future generations (Tilman et al., 2002). Since phosphorus in agriculture is the second most growth limiting macronutrient after nitrogen, its proper management in soil contributes significantly to sustainable crop production. In such soils where yield is limited because of inherent low P concentration (P deficient soils), application of relatively higher amount of mineral P fertilizers is the only way to enhance soil available P status to a target value in a long run that can sustain high crop yield. However, once the target value is reached, the available soil phosphorus concentration can be kept at a level that can sustain high crop yield through maintenance fertilization (replacing only the P removed from the field along with the harvested crops).

The P contained in crop residues left in the field can be recycled by incorporating the residues into the soil whereas part of P in crop residues fed to livestock can be returned back to the soil in the form of manure and also as bone meal. The mineralization of such organic P sources can occur through the action of microorganisms and plants exuding phosphatases and phytases. However, the P removed along with cereal grains, other edible vegetable parts and livestock products such as cow dung, milk and meat used for human consumption need to be replaced through mineral P fertilizer application. Therefore, under condition where P removed from the soil by harvested crop can be returned as crop residues and manures, the amount of mineral P fertilizer required for maintenance fertilization becomes less. In a nutshell, regular application of maintenance P fertilizers, incorporation of crop residues and application of organic manures can reduce nutrient mining and contribute to sustainable crop production.

Problems of mineral p fertilization and suggested management options The major portion (80-90%) of mineral P fertilizers applied to the soil can not be absorbed by plants due to adsorption to Fe oxides/ hydroxides, Al hydroxides as well as to Ca and Mg carbonate surfaces and due to chemical precipitation. This results in the formation of sparingly soluble Fe-phosphates (strengite) and Al-phosphates (variscite) in acid and Ca-phosphates in alkaline soils. Moreover, the applied mineral P fertilizer may also possibly be transformed to organic form, a process known as microbial immobilization (Holford, 1997). Thus, the mineral P fertilizer recovery of crops during the year of application is usually very low (less than 20%).

Mineral P fertilizer recovery of crops can be improved through proper method of P fertilizers application. In soils that have a high P fixing capacity, band application (where P is concentrated in a narrow zone) enhances P fertilizer recovery compared to broadcasting, since concentrating P in a small soil volume saturates the P binding sites and lowers the buffering capacity of the soil thereby increasing mobility of phosphate ion into the soil solution and its further diffusion toward the root. Band application especially of mono-and diammonium phosphate fertilizers enhance root proliferation due to both N and P effect, consequently improving the P uptake capacity of plants. Similarly, in alkaline soils having a high capacity to precipitate P as calcium phosphate, addition of organic materials such as farmyard manure along with mineral P fertilizers improve P solubilization through microbial activities such as excretion of organic anions, and H2CO3 formed due to respiration as well as H+ released by plant root induced by NH4+ uptake, all of which result in acidification of the rhizosphere. Besides rhizosphere acidification, the organic anions as well as the H2CO3 may also play role in desorption of phosphate ions adsorbed to Ca and Mg carbonates, through ligand exchange reactions. This enhances P availability contributing to improved mineral P fertilizer recovery by crops.

P-efficient crops/cultivars and their role in sustainable crop production Phosphorus efficiency, which is the ability of a crops/ cultivar to produce high yield under P limiting condition (Graham, 1984), can be attained through improved P uptake efficiency (the ability to take more P from the soil under P limiting condition) and/ or through improved P utilization efficiency (the ability to produce higher dry matter yield per unit of P taken up) (Gahoonia and Nielsen, 1996). Thus, P-efficient cultivars produce reasonably high yield in low P soils through either ways and thus can reduce mineral P fertilizer input requirement in agricultural production.

Phosphorus uptake efficient cultivars may contribute to sustainable crop production by producing reasonably high yield under P deficient condition due to their ability to exploit greater soil volume for accessing more P through producing larger root system (higher root-shoot ratio), longer root hairs or via forming association with mycorrhiza. Such cultivars may also enhance the applied mineral P fertilizers recovery and improve P availability, since they may be adapted to mobilize mineral P fertilizers fixed by the soil after application through exuding organic anions and protons.

Additionally, P uptake efficient cultivars may also be able to mineralize organic P sources (including those of plant and microbial origin) by releasing acid phosphatases, phytases and/or RNase, thereby increasing soil available P to sustain high yield. Thus, P uptake efficient cultivars are able to produce high yield at relatively low soil P status which can be reached by applying less amount of mineral P fertilizer.

On the other hand, P utilization efficient cultivars produce high yield per unit of absorbed P under P deficient conditions, since they have low internal P demand for normal metabolic activities and growth and hence have low requirement for mineral P fertilizer inputs to produce reasonably high yield. Moreover, they remove less P from soil during growth and therefore the quantity of P removed along with the harvestable parts of the crop would obviously be less, consequently reducing the quantity of mineral P fertilizer inputs required for maintenance fertilization.

Mechanism of Phosphorus Efficiency in Plants

Phosphorus efficiency is a term that generally describes the ability of crop species/genotypes of a given plant species to give higher yield under P-limiting condition. Plant species as well as genotypes within the same species may differ in P efficiency. The ability of a crop/ genotype to give higher yield under P-limiting condition may be related to: the ability to take up more P from the soil under P-limiting condition (uptake efficiency) or the ability to produce higher dry matter per unit of P in the plant tissue (utilization efficiency) or a combination of both.

6.1 P Uptake Efficiency

For increased P uptake efficiency, plant species/genotypes may use various adaptation mechanisms to gain access to previously unavailable soil P reserves such as through altered root morphology, exudation of chemical compounds into the rhizosphere and association of roots with mycorrhiza which are discussed throughout 5.1.1 to 5.1.4 of the next section. Higher P uptake efficiency is usually related to either larger root system size (usually higher root-shoot ratio) or to higher uptake rate per unit of root length.

Root morphology Root architecture: Root architecture refers to the complexity of root system spatial configurations that arise in response to soil conditions (Vance et al, 2003). Some plant species/ genotypes alter the architecture of their root systems under P stress conditions to optimize P acquisition (Richardson et al, 2011). Adaptations that enhance acquisition of P from the topsoil are important because of the relative immobility of P in the soil, with the highest concentrations usually found in the topsoil and little movement of P into the lower soil profiles (Vance et al, 2003). Studies with Phaseolus indicated that genotypes that have highly branched root systems and more root apices are efficient in acquiring P. Additionally, P-efficient genotypes also grow lateral roots from the basal roots at an angle that enables them to better explore the upper layers of the soil relatively rich in P (Lynch and Brown, 2001). Efficient genotypes develop an architecture that places active roots in regions of the soil more likely to contain available P (Smith, 2001).

Root-shoot ratio: Because of low mobility of phosphorus in the soil, some plant species/genotypes develop larger root systems that allow a plant to have access to greater soil volume so that higher quantity of soil P can reach the root surface for uptake (Jungk, 2001).

Higher root-shoot ratio is often reported for P stressed plants as compared with P sufficient plants (Gaume et al., 2001). This is due to severely reduced leaf growth under P stress, which leads to diminished leaf demand for assimilates consequently causing translocation of photosynthates to the root (Cakmak , 1994) for better root growth. Preferential root growth thus helps the stressed plants to acquire more P from the ambient environment in response to P stress conditions. Difference in P uptake efficiency between crop species (Fohse et al, 1988) and genotypes (Schenk, 2006) was noticed, which was accounted to difference in root-shoot ratio.

Root hairs: Root surface area alone may not be adequate to feed plants, especially with nutrient of low mobility like phosphorus. The presence of root hairs are therefore, quite important for the acquisition of poorly mobile nutrients such as P (Jungk, 2001). Root hairs are tubular outgrowths on the root surface and their formation is mediated by ethylene production (Michael, 2001). Root hairs substantially increase the root surface area for ion uptake. Root hairs have a smaller diameter than roots and grow perpendicular to the root axis, which allows better exploration of soil due to enhanced absorptive surface area. Root hairs form as much as 77% of the root surface area of field crops (Parker et al., 2000). Root hairs are also effective in extending the width of the P depletion zone around the root by increasing the volume of the soil explored for phosphorus.

Some plant species/genotypes are adapted to produce longer and more root hairs under P deficient conditions. However, due to overlapping of nutrient depletion zones developing around the root hairs it is not the density but rather the root hair length which is generally considered to be the most important attribute. Gahoonia and Nielsen (2004a) demonstrated that a barley genotype with a capacity to form longer root hairs (about 1 mm) took up more P, and tended to yield better when P was limiting crop growth compared to genotypes having root hairs half the length (0.5 mm). Gahoonia and Nielsen (1998) reported that root hairs contribute up to 63% of the total P uptake under P deficient condition. Thus, plant species or genotypes of the same species with different root hair length may exhibit different P uptake efficiency (Eticha and Schenk, 2001). However, unlike in soil culture, root hairs play no significant role in P acquisition under hydroponic culture, since nutrient diffusion is not a problem under such conditions (Bates and Lynch, 2000). Root hair growth is genetically controlled and thus traits conferring increased root hair length can be utilized in plant breeding programmes.

Root radius: Under P deficient condition, plant species/genotypes produce fine roots that allow a contact of larger soil volume per unit of root surface area, thereby increasing P uptake rates (Fohse et al., 1991; Gahoonia and Nielsen, 2004b). Thus, plant species/genotypes with thinner roots may be more effective in absorbing soil phosphorus (Gahoonia and Nielsen, 2004b). However, since fine roots tend to turnover more rapidly than coarse roots, the carbon cost of producing finer roots may be higher as these will have to be replaced more frequently (Sattelmacher et al, 1994; Gahoonia and Nielsen, 2004b). As a result of this, less consideration is given to select genotypes based on this root morphological trait and to use this trait in breeding programme.

Cluster rootformation: Proteoid and dauciform root clusters commonly occurring in plant species belonging to Proteaceae, Retinoaceae, Cyperaceae, Fabaceae and few other families are induced by P deficiency and they are adaptive mechanism to maximize P acquisition from unavailable P resources. Both proteoid and dauciform roots are covered with dense mat of root hairs, which markedly increase the surface area of the root system and are also specialized in efficient synthesis and secretion of organic anions (especially citrate and malate) and phosphatases, which help to solubilize insoluble P resources and hydrolyze organic P for plant uptake (Playsted et al, 2006; Lambers et al., 2006). Cluster roots grown under P deficiency exude 20-to 40-times more citrate and malate than those grown under sufficient P (Vance et al., 2003).

Association of roots with Arbuscular Mycorrhizae: The vast majority (82%) of higher plant species have the capacity to form a symbiotic association with mycorrhizal fungi (Brundrett, 2002). The degree of dependency on mycorrhizal association under P stress could differ with crop species those lacking root hairs such as onion being more dependent on the association (Deressa and Schenk, 2008) and also with cultivars (Zhu et al., 2001). The symbiotic association of plant roots with arbuscular mycorrhiza (AM) enhances the uptake of nutrients with low mobility like P especially when the species has a root system that is relatively coarse with few root hairs. A significant contribution of AM fungi to plant P uptake has been reported especially for soils with low P content and with high P fixing capacity (Marschner and Dell, 1994).

Increased P absorption by mycorrhizal hyphae is related to both increased physical exploration of the soil and modification of the root environment (Bolan, 1991; Smith and Read, 1997; Tinker and Nye, 2000). Enhanced nutrient absorption through physical exploration of the soil due to mycorrhiza is related to large length of hyphae per unit root length, smaller hyphae radius and larger surface area (Tinker et al, 1992). Particularly the ability of hyphae to extend several centimeters out into the surrounding soil allows to expand the effective volume of the soil that the plant can exploit (Smith, 2002). Arbuscular mycorrhiza fungi also hydrolyze organic phosphate through releasing acid phosphatase into the soil thereby contributing to increased P uptake of the host plant (Tawaraya et al., 2005). Arbuscular mycorrhiza hyphae store polyphosphates in their vacuoles, which may be hydrolyzed in the arbuscules and transported as inorganic P into the host plant across the plasma membrane of cells (Smith and Gianninazi-Pearson, 1988). The contribution of AM fungi to P uptake reach up to 77% under low P supply compared to only 49% under high P supply (Thingstrup et al., 2000). Furthermore, Deressa and Schenk (2008) reported that fungal hyphae accounted for nearly the whole of predicted P uptake by onion.

Root Exudation

Organic Acids

Plants growing in an ecosystem low in available P have to obtain P from adsorbed P, sparingly soluble P and organic P complexes. Many plants have developed elegant biochemical mechanisms to solubilize P from insoluble P complexes thereby increasing the pool of P available for uptake (Raghothama and Karthikeyan, 2005). Organic anions such as citrate and malate are the major root exudates released, in response to P deficiency for mobilizing P for plant uptake (Neumann and Romheld, 1999; Dechassa and Schenk, 2004). The range of organic anions released is, however, dependent on the plant species (Ohwaki and Hirata, 1992; Dechassa and Schenk, 2004). Genotypes of the same species may also differ in their ability to exude organic anions and hence in their ability to mobilize P from sparingly soluble P sources (Dong et al., 2004; Corrales et al., 2007). Organic anions mobilize inorganic P through complexing metal cations that bind phosphate and displace phosphate from the soil matrix by ligand exchange (Raghothama and Karthikeyan, 2005). Besides its role in P solubilization, organic anions exudation also protect the roots of some plants from Al toxicity (Ma et al., 2000) thereby enabling root proliferation and increased foraging capacity for P in acid soils (Smith, 2001).

Acid phosphatase and phytase: A major portion (30-80%) of total P in soil is present in organic forms (Vance et al., 2003). Half of this is in the form of phytin and its derivatives (Tarafdar and Claassen, 1988). This organic-P complex needs to be hydrolysed by enzymatic activities before the inorganic P is released into the rhizosphere for plant uptake. Acid phosphatase and phytase are a group of enzymes produced by plants in response to P stress that can hydrolyse a range of organic-P forms thereby enhancing plant P uptake from unavailable P resources. Acid phosphatases functions both as intracellular (vacuolar) and extracellular (secreted) P salvage systems that catalyze the hydrolysis of P from phosphate-monoesters. The intracellular acid phosphatases play significant role in remobilizing P during senescence and P stress (Plaxton, 2004). Phytases secreted by plant roots into the soil hydrolyze the inositol-phosphate while the intracellular phytase will degrade phytic acid which is the principal storage form of phosphorus in seeds and pollen for remobilization and use during seedling growth and pollen germination (Li et al., 1997).

The amount of phytases and acid phosphatases secreted in response to P deficiency differ between plant species as observed by Li et al. (1997). Increased exudation of acid phosphatases was observed in several maize genotypes, except with one genotype under P deficiency. This indicates that genotypes may also differ in acid phosphatase activity under P stress (Gaume et al., 2001). Similarly, barley genotypes exhibited different extracellular phytase activity under P deficiency (Asmar, 1997). Moreover, secretion of RNase degrades the nucleic acids present in decaying organic matter (Plaxton, 2004). Thus, RNase represents an important source of extracellular P that is exploited by P stressed plants. Some plant species/genotypes also release protons into the soil to acidify the rhizosphere condition to enhance P uptake from acid-soluble Ca-phosphate (Neumann and Romheld, 1999).

Others: Induction of high-affinity Pi transporters in roots of P deficient plants also play a crucial role in the acquisition of limited P by some plants (Neumann and Romheld, 1999; Plaxton, 2004). Enhanced expression of high-affinity, plasma membrane-bound Pi transporters in roots and a concomitantly increased P-up-take capacity, was reported as a typical P-starvation response (Dong et al, 1999). However, other reports indicated that diffusion of P in the soil is the key limiting factor for P uptake and a change in uptake systems have little effect on the plants capacity to acquire P from the soil (Raghothama and Karthikeyan, 2005).

P utilization efficiency: Besides increased acquisition of soil P, efficient utilization of acquired P is also considered an important adaptation for plant growth on low P soils. Phosphorus utilization efficiency refers to the ability of a plant species/genotype to produce higher dry matter per unit of P absorbed.

The mechanism of higher internal P utilization efficiency is not clearly known. However, it may be related to the ability of a plant in releasing inorganic P from the storage pool (vacuole) to the cytoplasm (cytoplasmic P homeostasis) or to selective allocation of P between cytoplasm and vacuole in favour of cytoplasm thereby ensuring sufficient Pi concentration in metabolically active compartments for normal functioning of plant metabolism (Lauer et al., 1989a; Raghothama, 1999). Additionally, higher internal P utilization efficiency may also be due to lower metabolic requirement for inorganic P at cellular level under P stress possibly due to the presence of alternative P-independent enzymes/metabolic pathways and/or energy sources.

Cytoplasmic P homeostasis: The fastest and largest manifestation of P starvation is a decline in intracellular P (Pi) concentration (Natr, 1992). However, plants are still able to maintain cytoplasmic Pi either through effective buffering with vacuolar Pi or possibly through selective allocation of Pi between cytoplasm and vacuole to constantly keep sufficient Pi in metablically active compartment (cytoplasm) despite the P stress as confirmed by 31P-nuclear magnetic resonance (NMR) studies (Lauer et al., 1989a). During nutritional P limitation, vacuolar Pi is released into the cytoplasm in a regulated manner that correlates with the severity of the Pi stress, ultimately ensuring a relatively constant Pi concentration in the cytoplasm (Mimura et al., 1996). The efficiency of this process is, however, dependent on the relative permeability of the tonoplast to Pi (Schachtman et al., 1998) and may vary between different plant species. Thus, the decline in cytoplasmic Pi, due to absence of effective Pi homeostasis directly affects sugar-phosphate export from the chloroplast (Flügge et al., 1980). This situation leads to the decline in Pi levels in the chloroplast stroma and an increase in starch synthesis. The decline in the stromal Pi can limit the photosynthetic capacity and hence plant growth rate.

There are divergent reports as to how P deficiency results in non-stomatal induced decline of plant photosynthetic capacity. Most common reports state that P deficiency reduces photosynthetic capacity through: (i) directly affceting ATP production (Rao et al., 1989), (ii). inactivation of enzymes involved in the ribulose bisphosphate (RuBP) regeneration (Fredeen et al., 1990) (iii) inactivating RuBP carboxylase enzyme which catalyzes CO2 fixation (Lauer et al., 1989b), (iv) combined effect of ii and iii (Brooks et al., 1988). High P utilization efficiency (high yield or relative growth rate per unit of P) of a plant species/ genotypes under P deficiency can be related to higher net carbon fixation (Yongfu et al., 2006) achieved possibly through effective cytoplasmic P homeostasis or through selectively allocating more Pi to the cytoplasm. Higher P utilization efficiency under P limiting condition may also be caused due to lower carbon demand for root respiration (Nielsen et al., 2001).

The use of P-independent enzymes/pathways in metabolismIn general, little is known about metabolic adaptations of plant respiratory pathways to P deficiency. Alternative use of P-independent enzymes instead of P-dependent ones in glycolysis pathways have been reported as plant respiratory adaptation to P deficiency.

Moreover, the P-independent glycolytic enzymes may also facilitate interacellular Pi recycling, since Pi is the bi-product of the reactions catalyzed by these enzymes, which can be reassimilated into the metabolism of P stressed plants (Plaxton, 2004). Besides the use of alternative P-independent enzymes, the use of alternative energy source such as the use of pyrophosphate (PPi) in stead of ATP have also been noticed in glycolytic pathways under P stress (Duff et al., 1989; Plaxton and Carswell, 1999). Pyrophosphate is a by-product of anabolic reactions including DNA, RNA, proteins, lipids and polysaccharide biosynthesis (Plaxton and Carswell, 1999). Plant cytosolic PPi levels are remarkebly insensitive to abiotic stresses such as P-starvation, which elicit significant reductions in cellular ATP pools. The large amount of PPi produced during biosynthesis may be employed by plants to enhance the energy efficiency of several cytosolic processes (Plaxton, 2004). Active transport of protons into the vacuole by PPi-dependent H+ pump in the tonoplast can use PPi instead of ATP as an energy donor, which can replaces the limited ATP pool under P deficiency. Plants may also use alternative mitochondrial respiration such as nonphosphorylative pathways that can bypass energy-requiring sites. This allows continued functioning of the citric acid cycle and respiratory electron transport chain under limited ATP production due to P stress.

Maintenance of cell-division and epidermal cell expansion One of the most striking effect of P deficiency is a reduction in leaf growth both in terms of leaf number and individual leaf size (Lynch et al, 1991; Chiera et al., 2002). Leaf initiation at the shoot meristem and its lateral expansion are controlled by the activities of cell division (Chiera et al., 2002; Assuero et al., 2004). Pi was reported to play a significant role in cell division activities (Sano et al., 1999). However, plants may still differ in tolerance of P stress effect on leave growth mainly due to difference in the ability to maintain cell division (Chiera et al, 2002; Assuero et al., 2004), leaf epidermal cell expansion (Radin and Eidenbock, 1984) or both (Kavanova et al., 2006), under lower tissue P concentration. Alternatively, plants may also retranslocate limited P from older leaves to younger leaves to maintain Pi at levels that permit optimal physiological activities including cell division (Plaxton and Carswell, 1999; Akhtar et al., 2008). Other possible mechanism of efficient P utilization includes conservation of Pi by replacing membrane phospholipids with non-phosphorus galacto and sulfonyl lipids (Plaxton, 2004).

How do we determine how much phosphorus to add- The answer is that we must account for the P-fixation capacity of the specific soil. For some Hawaii soils, researchers have determined the P-fixation capacity as various levels of phosphorus is added to the soil. This information allows us to predict how much phosphorus must be added to the soil to achieve a target phosphorus level. To view the P-sorption curves for selected soils, click on the link below:

PHOSPHORUS LEACHING AND RUNOFF: In Maui County and other tropical regions, highly weathered soils often provide little available phosphorus for plant growth. To further compound this issue, agricultural systems can experience phosphorus losses as the result of erosion by wind and runoff water. Erosion by wind can carry particles that contain sorbed-P to water systems, where phosphorus may later desorb. Sediments containing phosphorus can also contaminate ground and/or surface waters. Additionally, phosphorus availability is reduced by the removal of plant material (which can serve as a source of organic phosphorus) during harvest.Although phosphorus leaching is normally limited in most Hawaii soils due to their high P-fixing characteristics, phosphorus leaching can occur if the soil reaches maximum phosphorus holding capacity, especially when P fertilizers are overapplied. Sandy soils are most susceptible to phosphorus leaching. The consequence of phosphorus leaching is the contamination of ground water reserves.

Chapter 5

Potassium: Role, Dynamics and Management

Introduction

Potassium (K) is a macronutrient required by all plant and animal life. While K is not a commonly limiting soil nutrient in crop production in Alberta, about 15% of Alberta soils used for annual crop production are estimated to have slight to moderate K deficiency.

Adequate soil potassium results in:

- superior whole-plant quality due to improved photosynthesis efficiency
- increased crop resistance to some diseases
- greater water use efficiency
- a normal balance between carbohydrates and proteins
- stronger straw and improved seed filling for cereal crops

Potassium deficiency in cereal crops results in:

- reduced growth
- delayed maturity
- lodging caused by weaker straw
- lower bushel weight

Potassium is Required by all Plant and Animal Life

Soil Potassium

Potassium is an indispensable nutrient for plant growth and is classified as a macronutrient because plants take up large quantities of it. A plant is only able to absorb potassium from the soil in a soluble form (K+). Potassium is highly mobile in the plant and plays a significant role in water management. Contrary to most other nutrients, potassium in the plant is not converted into complex organic molecules (except wood), but acts as a critical co-factor in facilitating numerous processes:

- Regulation of water pressure (turgor) in plant cells, important for stretching the cell walls, respiration (opening and closing of stomata), water intake and the orientation of leaves towards light.
- Activation of enzymes.
- Build up and transport of organic compounds in the plant such as sugars, amino acids and starch.
- Regulation of pH in plant cells.
- Capture of carbon dioxide during the photosynthesis process.
- Balancing of electrical load in various parts of the plant cell.
- When a crop is able to absorb sufficient potassium it has a greater ability to withstand stress and damage caused by insects and/or diseases.
- Potassium requirement of the crop during the growing season

For plants, potassium is the second more important constituent after nitrogen. A reduced potassium availability will cause growth to slow and the plant will extract potassium from old leaves. A severe potassium deficiency first manifests as chlorosis (yellowish discolouration) in the lower parts of the plant and especially in the tips and edges (tip burn). In addition, it impairs the lignification of vascular bundles, which causes plants to weaken and become susceptible to lodging.

The plant's need to uptake potassium sharply increases during the season. There is a relatively high demand for potassium particularly during the period of flower formation and fruit set. As a positively charged ion, potassium is very important for the uptake and transport of negatively charged ions such as nitrate, phosphate and amino acids. Potassium plays a vital role in the translocation of sugars and starch from the leaves through to storage organs such as fruits, bulbs and tubers. If a potassium deficiency occurs during this period, it affects the quality of the produce to be harvested. In fruit vegetables, potatoes and onions in particular, potassium plays a crucial role in the ripening and storage phases.

The majority of soils in Alberta contain sufficient plant-available potassium to satisfy crop growth. The total amount of K in soil often exceeds 40,000 kg/ha (36,000 pounds per acre - lbs/ac) in the top 15 cm (6 in) of soil. However, only 1 to 2% of the total K in soil is in a form available to plants. The parent geologic material on which Alberta soils developed generally contains considerable K-bearing clay minerals. Potassium becomes available to crops through weathering of these soil minerals.

Forms of K in Soil

There are 3 forms or pools of inorganic K in soil:

1. Unavailable K About 90 to 95% of the total soil K is locked within the structure of the layered clay mineral sheets and is not available to plants.
2. Slowly available K About 5 to 10% of the total soil K is slowly becoming available to plants. Weathering of the clay minerals occurs on the surface of the minerals and results in a very slow release of K from the unavailable K pool locked within the clay minerals. The weathering of clay minerals gradually releases K from the minerals to recharge the K removed from the available soil K pool.
3. Available and exchangeable K Available and exchangeable K represents about 1 to 2% of the total soil K. The K in soil available to plants is dissolved in soil water while exchangeable K is loosely held on the exchange sites on the surface of clay particles. Typically, this K pool or fraction represents about 1 to 2% of the total soil K. A portion of this pool is plant-available K dissolved in the soil water.

The exchangeable K, which is positively charged (K+), is loosely held on the negatively charged exchange sites on the surface of clay minerals and is referred to as exchangeable K. As the available K dissolved in the soil water is taken up by plant roots, exchangeable K is released into the soil solution to maintain an equilibrium between the 2 forms. Soil tests attempt to measure the available and exchangeable K in soil to determine the K-supplying power for the soil K for crop production.

Available and exchangeable levels of K generally range between 300 and 1,000 kg/ha (270 to 900 lbs/ac) in Alberta soils in the top 15 cm (6 in) of soil. A very small percentage of Alberta soils have as little as 100 kg/ha (90 lbs/ac) of available K. A minimum of 200 kg K/ha

(180 lbs/ac) in the top 15 cm (6 inches) of soil is generally required for adequate growth of most crops grown in Alberta.

K adsorption: Potassium only occurs in soils in inorganic form and does not make up part of the soil organic matter. Potassium in soil solution and in exchangeable form occurs as a positively charged ion, K+. Soil colloids and the surface of soil clay minerals are negatively charged. Therefore, K will tend to be adsorbed onto the surface of negatively charged soil particles. As a result, soil K tends to be fairly immobile in soil and is not subject to leaching or downward movement in soil. Soils that have the greatest potential for K deficiencies are the coarse (sandy) and medium (loam) textured soils in the Black, Gray-Black and Gray soil zones as well as the organic soils in Alberta. Irrigated coarse textured soils that have intensive crop rotations that include potatoes, sugar beets and/or alfalfa in the rotation may require K. A soil test is often very useful to determine whether K fertilizer application is warranted.

Plant K uptake: It is important to note that plants take up K in ionic form as K+. Potassium fertilizer is sold as K2O. The most commonly sold K fertilizer in Alberta is K chloride (KCl) (0-0-60), which contains 60% K2O. To convert from lb of K2O to lb of K, divide by 1.6. Soil tests normally report available and exchangeable K as K and provide K fertilizer recommendations in lb of K2O.

Potassium deficiency: Potassium deficiency symptoms on cereal grains appear as a burning or scorching of the older, lower leaves. The burning begins at the leaf tip and continues down the leaf margin. Lower leaves are affected first because K in these older leaves is translocated to the new upper leaves to meet growth requirements.

In alfalfa, K deficiency appears as white or yellow spots on the lower leaves. Many other factors can affect the appearance of lower leaves, so a diagnosis based on visual deficiency symptoms must be confirmed with soil tests. Potassium deficiency in Alberta is less common than nitrogen, phosphorus or sulphur deficiency for 3 reasons:

1. The parent geologic material on which the soils developed generally contains considerable K-bearing clay minerals.
2. Alberta soils are young with minimal weathering and leaching.
3. Crop removal has been relatively minor where cereal, oilseed and pulse crops have been the dominant crops, especially where straw was left on the field or manure returned. About 80% of the K taken up by crops remains in the straw and is returned to the soil.

There are an estimated 3,000,000 acres of K-deficient soils in Alberta. Of this total:

- 2,500,000 are moderately deficient
- 500,000 are very deficient

K-deficient soils tend to be sandy (light to medium textured), alkaline, carbonated and/or imperfectly to poorly drained in their natural state. Organic soils are also frequently deficient in K.

Forage, silage, potato and sugar beet crops have a high K requirement; they take up and remove relatively large amounts of K from soils. When these crops are included in an intensive crop rotation on sandy soils, there is an increased risk of K deficiency.

Potassium is a relatively inexpensive nutrient, and deficiencies can often be corrected with moderate rates of application.

Correcting a K deficiency can result in excellent economic returns. Potassium chloride (KCl) (0-0-60), also called muriate of potash, is the most commonly-used K fertilizer, which contains 60% K2O.

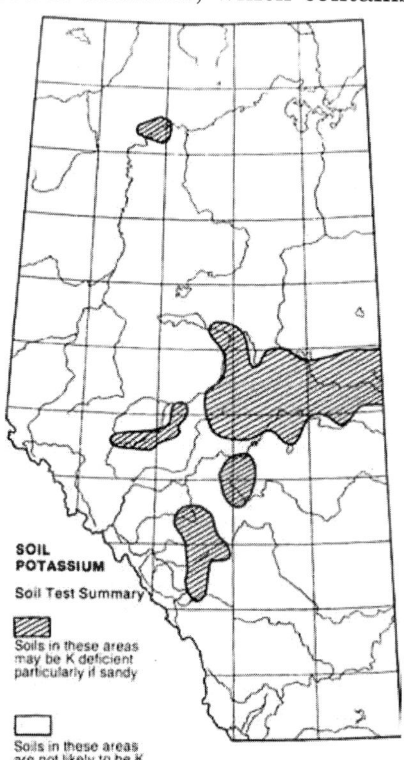

Figure: *Soil areas of Alberta more susceptible to having K-deficient soils.*

Table 1. *Approximate nutrient uptake of crops for specific yields*

	Yield			N		P_2O_5		K_2O	
		grain	straw	grain	straw	grain	straw	grain	straw
	(bu/ac)	(kg/ha)		(lbs/ac)					
Wheat	40	2,400	3,600	60	25	25	6	15	70
Barley	80	3,400	3,500	70	30	35	10	25	90
Canola	35	2,000	3,000	65	40	35	15	20	70
Flax	25	1,500	2,300	50	20	20	6	15	45
Tubers									
Potatoes	50,000 (20 t/ac)			140		65		300	
Hay									
Alfalfa	11,000 (5 t/ac)			280		65		300	
Grass	7,000 (3 t/ac)			100		30		130	

Crop K requirements: The potassium content of the above-ground portion of commonly grown crops is shown in Table 1 above. Note the distribution of K between grain and straw as compared to nitrogen and phosphorus. Potassium removal from the soil is relatively low when only the grain of cereals is removed from the field. Potassium removal is much higher when forage, silage crops or potatoes are grown. If the forage is fed to livestock on the farm, the manure can be used to return much of the K removed by the forage crops.

While an awareness of the amount of K removed by various crops is useful, it is not in itself a basis for determining fertilizer requirements. The majority of soils in the prairie region contain high levels of plant-available K and can supply adequate amounts, even for high-use crops, for many years. On soils marginal to deficient in K, high-use crops will require higher rates of K fertilizer to maintain adequate nutrition.

Determining potassium fertilization need: The easiest way to determine the need for K fertilization is through a soil test. In Alberta, response to K fertilizer has been related to the amount of K extracted from the soil with ammonium acetate. The most common soil test method used in Alberta to extract K from the soil is the modified Kelowna method.

In studies, large increases in barley yield were usually obtained when K fertilizer was applied to soils with less than 151 lbs/ac of extractable K. Sites with a soil test K level of less than 150 lbs/ac all responded to added K fertilizer. On soils with 200 to 250 lbs/ac of extractable K, moderate fertilization of 15 to 30 lbs/ac of K2O may result in a profitable response.

The probability of response above a soil test level of 250 lbs/ac was low. However, at soil test K levels between 250 and 300 lbs/ac, producers should consider a maintenance application of 15 to 20 lb K2O/acre to help replace K taken up by crops. This application could also potentially improve crop yield in areas of a field that may have slightly deficient soil K levels.

A response to K fertilization is sometimes obtained on soils not considered deficient in K. Research, principally in Oregon, Washington and South Dakota, has shown that the presence of chloride in K chloride fertilizer can result in increased yield through the suppression of plant diseases such as take-all and common root rot in wheat and barley. Such responses have not been observed in Alberta research with wheat and barley; therefore, the use of K chloride on non K-deficient soil for the suppression of disease must be on a trial basis.

Research in Montana has indicated that K deficiency may occasionally occur on soil with high soil test K because of slow K diffusion in cold, dense soils. These results have been used to promote the need for K fertilization for early seeding of cereal crops in central and northern Alberta.

Research in Alberta on K-deficient soils has shown equal response of barley to K with early and later seeding. If soil K was less available at lower soil temperatures, greater response to K should have been obtained with early versus later seeding. The results do not support a higher K requirement for early seeded cereal crops.

Optimal potassium fertilising: Supplying minerals in the form of animal manure, compost and/or fertiliser will top up the levels of potassium in the soil at the beginning of the season. However, the availability of potassium greatly depends on the climatic conditions. Potassium can only be taken up the plant when it is sufficiently dissolved in the moisture in the soil. Too wet conditions will cause potassium to leach out relatively easily (especially on sandy soils) and relatively dry conditions reduce availability. This makes applying extra potassium during the season in the form of a foliar fertiliser a very interesting option. Foliar fertiliser offers many advantages, especially in crops such as potatoes, sugar beets and onions that have a high potassium demand. Especially for these applications, Soiltech has formulated the product 'Optima Leaf-K'. This consists of organically complexed potassium. Unlike other potassium fertilisers, Optima Leaf-K has a relatively low EC (salinity), a neutral pH and has good miscibility with other substances. The product has proven its value multiple times in field trials in recent years, including with onion crops. We are pleased to show you the results of a field trial with onions. These figures were achieved in the extremely dry summer of 2018.

Treatment with Optima Leaf-K from Soiltech resulted in a +9% higher net yield in the trial. A very clear difference was noticeable, particularly in sizes >40 cm.

Treatments	Specifications	Kg K2O
Control	Standard fertiliser (240 kg KAS per ha)	–
Kali-60	200 kg/ha	120 kg/ha
Patent-Kali	400 kg/ha	120 kg/ha
Powerleaf Kali	3 x 3 liter/ha	3,4 kg/ha
Bladkali TS	3 x 5 liter/ha	6,5 kg/ha
Optima Leaf-K	3 x 5 liter/ha	5,2 kg/ha

Choosing application rates: Application rates are based on the soil test level and the responsiveness of the crop. On soils marginal to deficient in K, crop removal (Table 1) should also be taken into account when planning an on-going K fertilization program.

Rates above 15 to 30 lb of K2O/ac for cereals and pulse crops should be banded or broadcast to avoid seedling injury. At low rates of application, placement with the seed is more effective than banding, and banding is more effective than broadcast. See Application methods below. The majority of K-deficient soils in Alberta are only slightly to moderately deficient. Maximum response to K on these soils is usually obtained with an application of 15 to 30 lbs/ac of K2O to cereal crops. On soils that are very deficient in K (soil test K less than 150 lbs/ac), high rates of K fertilization are usually required to achieve maximum productivity. On very K-deficient soils, soil K can be built-up with an initial application of 200 to 300 lbs/ac of K2O (broadcast and incorporated), followed in subsequent years with annual applications of about 30 lbs/ac. Similar results could be achieved by banding 50 to 80 lbs/ac for a few years to build up soil K levels.

Research in Alberta indicates that soil tests do not always adequately reflect residual fertilizer K where high rates have been applied to very K-deficient soils. Following the application of up to 810 lbs/ac of K2O on very K-deficient soils, soil test K was still in the deficient range, but subsequent crops showed little or no response to additional K fertilizer. These heavily fertilized soils contained residual, crop-available K that was not reflected in the soil test.

Therefore, when high rates of K fertilizer are applied to very K-deficient soils, the subsequent need for additional fertilizer should be determined on the basis of crop response, rather than solely on the basis of a soil test.

Types of K fertilizer: The most common form of potassium fertilizer used in Alberta is potassium chloride (KCl), which has the analysis 0-0-60 or 0-0-62 (Table 2). It is mined and refined in Saskatchewan. Potassium chloride can be safely blended with nitrogen and phosphate fertilizers to produce various blends to achieve optimum N-P2O5-K2O application.

Table 2. Potassium fertilizers available in Alberta

Nutrients -% by weight					
Name	**N**	**P_2O_5**	**K_2O**	**S**	**Remarks**
Potassium chloride	0	0	60	0	Are the most commonly available K fertilizers and least expensive
	0	0	62	0	
Potassium sulphate	0	0	50	18	Contains sulfur as well as K

Application Methods

Annual Crops

Since soil potassium is not mobile in soil, placement of K fertilizers with or near the seed is usually the most effective and efficient method of application provided the rate of application is not greater than the seed can tolerate. Potassium chloride (KCl) is a salt. If too much K or a combination of other fertilizers that have a high salt index are placed with the seed, a "salt effect" can delay or reduce seed germination and emergence. The salt effect of fertilizer will interfere with the moisture uptake by the seed and may result in the death of a germinating seed.

The safe level of K that can be applied with the seed depends on the crop. In general, smaller seeded crops such as canola have a lower tolerance than cereal crops. The clay and organic matter content of the soil and the soil moisture content will also have an effect on possible germination problems. Soils with a higher clay content or higher organic matter content can hold more water. This situation can slightly reduce the salt effect when soil moisture levels are good by diluting the amount of K salt dissolved in the soil water.

With average soil moisture conditions and medium- textured soils, the total amount of seed-placed fertilizer materials should not exceed 175 lbs/ac, and the total amount of N plus K2O should not exceed 40 lbs/ac at a seed-bed utilization (SBU) of 10%. For less tolerant crops such as canola, flax and peas, the application of K with the seed should not exceed 15 lb K2O/acre, provided other fertilizers are not seed placed.

These recommendations are based on the use of a seed drill with a 10% SBU, which places the seed and fertilizer in a very narrow band. If the opener spreads the seed over a wider band, higher rates of fertilizer can be safely placed with the seed.

Side-banding: Side-band placement is an efficient means of applying K and much safer than seed placement, particularly when higher rates of K must be applied. Ideally, with this placement, the fertilizer is in a band approximately 2.5 cm (1 inch) to the side and possibly 2.5 cm (1 inch) beneath the seed. This separation of fertilizer and seed reduces the possible detrimental effects on germination when high rates are applied. There are a number of different openers that have slightly different placement positions.

Banding: Banding, also referred to as deep banding, places the K into the soil in a concentrated band prior to seeding. It is believed that this placement method should provide similar results to side-banding K.

Broadcasting: Broadcasting K before seeding is less efficient than applying K in a band with or near the seed. The major role for broadcast applications of K fertilizer will be in building up soils extremely deficient in K or for use with forage crops.

Perennial crops: For perennial forage crops, potassium is best applied by broadcasting and incorporating at higher rates of 100 lb K2O/acre before seeding. Higher rates should be used on K-deficient irrigated land. This approach will overcome the problem of limited movement of K into the soil when applications are made after stand establishment.

Where established stands require K fertilizer, broadcast applications are the only option, and relatively high rates may be required on severely K-deficient soils. Fall or spring applications could be made, but fall applications would likely be preferred in dry areas because of the additional winter moisture available to move the K into the root zone. For K-deficient soils, K fertilizer application in early fall may help reduce alfalfa winterkill and help maintain the proportion of alfalfa in mixed forage stands.

Calcium, Magnesium and Sulphur-Role, Dynamics and Management

Calcium and Magnesium

Plant uptake, cation adsorption and desorption in soil, leaching from rainfall and irrigation, and weathering of minerals all contribute to the concentration of water soluble Ca and Mg available to meet tree nutritional needs. Water soluble cations are determined from the saturated paste extract soil test procedure while the exchangeable cations are determined with the ammonium acetate procedure. Also important are the concentrations of exchangeable (non-water soluble) Ca and Mg which help to promote favourable soil structure. Soil chemistry is in a constant state of change attempting to reach equilibrium between the soluble and non-soluble (exchangeable and mineral) phases. The May 2009 newsletter discussed this dynamic process. Calcium and magnesium share similar chemical properties in soils. Both Ca and Mg are "double positively charged (divalent) cations in the soil-water phase and on soil cation exchange sites. Calcium is adsorbed to soil exchange sites preferentially and more strongly than Mg. When Ca and Mg are abundant in the soluble phase tree roots absorb these nutrients by mass flow.

If Ca and/or Mg are less abundant or limited by soil moisture, uptake occurs more slowly through diffusion. Table 1 provides ranges of exchangeable Ca and Mg levels that may be observed in soils in the Sacramento Valley. The ranges are expressed in three different reporting units that may be used by agricultural laboratories. The levels of exchangeable Ca and Mg can vary widely in soils and may correlate closely with levels of water soluble Ca and Mg (i.e. when exchangeable levels are high, water soluble levels are also usually high and conversely).

The amount of organic matter as well as the type and amount of clay largely determine the cation exchange capacity (CEC) of a soil. The cation exchange capacity can be thought of as a measure of the reservoir of Ca, Mg, K, and other cations adsorbed to the soil in the non-soluble phase that are available to replenish the nutrients taken up in the water soluble phase or may be absorbed directly by nearby roots. Soils which have a large amount of sand with very little clay or organic matter have little CEC or a reservoir to hold cationic nutrients. Adding large amounts of calcium, magnesium and potassium or even anions such as boron, sulfates or nitrates can be easily leached from sandy soils, those having low organic matter content or low CEC. Table 1. Common ranges in exchangeable calcium (Ca) and magnesium (Mg) in soils in California expressed in three different soil test reporting units. Element meq/100 g soil1 ppm (mg/kg)2 lbs/ac-ft soil3 Calcium (Ca) 5 - 50 1000 - 10,000 4000 - 40,000 Magnesium (Mg) 2 - 30 240 - 3600 960 - 14,400 1 The units of meq/100 g soil imply the determination of exchangeable cations with the ammonium acetate procedure. The units for saturated paste extract determinations are expressed as meq/L. 2 The units of meq/100 g soil are converted to ppm with the use of the equivalent weight of the cation. 3 The ppm concentration is multiplied by 4 using the assumption that one acre foot of soil weighs four million pounds and one acre 6 inches weighs two million pounds. Many California soils are alkaline (pH above 7.0) with an abundance of Ca and Mg bearing minerals in the soil. Under these conditions, Ca and/or Mg deficiencies are generally uncommon. In contrast, some orchard soils in the Sacramento Valley are acidic (pH 5.0 to 6.5) with a high concentration of hydrogen ions on the cation exchange sites causing growers to have more management questions about Ca and Mg levels in the soil. Most agricultural laboratories utilize two soil test methods to evaluate soil cations and their relationships.

The saturated paste extract procedure is used for salinity evaluation, particularly the Ca, Mg and Na concentrations. Evaluating soil salinity will be a topic in a future article. The ammonium acetate procedure is used to determine the exchangeable Ca, Mg, K (potassium) and Na (sodium) concentrations as well as the base saturation percentages. In general, field research has shown that plant growth is not affected over a wide range of Ca to Mg ratios, from 6:1 to 1:1. Either the saturated paste extract concentrations (meq/L) or the base saturation percentage (%) can be used to determine the Ca:Mg ratio. Magnesium deficiencies are likely to occur if Ca:Mg ratios are greater than 8:1. Unfavourable plant growth along with poor soil structure may occur when Mg concentrations are greater than twice those of Ca. Blossom end rot and white or pink centers in tomato is an example of a crop in the Sacramento Valley that may develop severe problems resulting from higher magnesium concentrations. Economic benefits from the application of gypsum as the most desired source of Ca to orchard soils when Ca:Mg ratios are in the range of 1:1 to 1:2 is highly unlikely. Base Saturation is estimated and reported by some but not all agricultural laboratories in California. It is an estimate of the proportions of Ca, Mg, K, Na, and H cations that occupy the cation exchange sites of a soil. Some laboratories also suggest desirable Base Saturation ranges of 60-80 percent for Ca and 10 to 20 percent for Mg. The concept of base saturation and the associated target levels stems from soil testing, and field observations conducted on acidic soils in the Central and Eastern U.S. that date back to the 1940's. Relationships between crop response and specific base saturation percentages of Ca and Mg levels or other cations in soils have not been helpful in improving crop yields or quality. Soil scientists have learned that more general ranges in the base saturation percentages are appropriate for soils. Even though some soils in the Sacramento Valley are acidic like soils in the southeastern and eastern U.S., they consist of different clay minerals that have much higher cation exchange capacities, which can lead to recommendations of extraordinarily high and uneconomical rates of Ca amendments to maintain the "Base Saturation" within the specific desirable ranges often suggested.

Calcium: Plants absorb calcium in the form of Ca2+. Following are the main functions of calcium in soil.

Functions of Calcium in plants

1) Calcium helps in the development of roots and growing buds of plants.
2) Calcium is essential for cell growth and cell wall formation.
3) Calcium makes the grains and stems hard in plants.
4) Calcium develops root nodules in leguminous crops.
5) Calcium is helpful in carbohydrate operation in plants.
6) Calcium neutralizes the effect of organic oxalic acids in plant metabolism.
7) In the presence of calcium, the absorption of nitrate by the plant roots increases and protein formation is increased.
8) Calcium helps in the growth of primary roots and root hairs in plants.

Deficiency symptoms of calcium in plants

1) Deficiency of calcium leads to drying of terminal buds.
2) Due to calcium deficiency, buds and fruits wither in immature stage.
3) Root development remains incomplete.
4) The edges of the new (upper) leaves are curled and wrinkled.
5) The margins and tip of the leaves may die.

Magnesium: Plants absorb magnesium in the form of Mg^{2+}.

Functions of Magnesium in plants

1) Magnesium is mainly helpful in the formation of chlorophyll in plants.
2) Helps in the translocation of nutrients inside the plants and absorption of nutrients in plants.
3) Many enzymatic activities inside the plant take place smoothly in the presence of magnesium, which helps in the formation of proteins, fats and carbohydrates.
4) Magnesium helps in phosphate metabolism and plant respiration.

Deficiency symptoms of magnesium in plants

1) Due to the high mobility of magnesium in plants, the symptoms of Mg deficiency first appear on older leaves.
2) Magnesium deficiency causes chlorosis in plants.
3) The colour of the lower (old) leaves of the plant is lost from the tip and margins and later the colour of the entire leaf is lost.
4) The veins remain green.
5) In cotton, the leaves can be light purple in colour with green veins.
6) Magnesium deficiency causes leaves to become smaller in size and curl upwards.
7) Deficiency of magnesium leads to fruit drop from trees.
8) Leaves of plants of lemon species become speckled.

Sulfur

Sulfur (S) undergoes numerous transformations in the soil involving biological, chemical, and atmospheric processes, similar to the nitrogen cycle. Plant uptake, mineralization, immobilization, soil anion exchange reactions, volatilization, leaching, oxidation and reduction, and mineral weathering all affect the availability of sulfur to trees. Soil properties such as water content, pH, temperature, and aeration also affect these processes. Trees absorb sulfur from soil in the water soluble, inorganic form of sulfate (SO_4^{2-}). Usually less than 10 percent of the total S in the soil is in the water soluble SO_4^{2-} form. Sulfate (SO_4^{2-}) taken up by the trees or lost to leaching is replenished by mineralization of organic matter, oxidation of reduced S, and other processes such as weathering of minerals that contain S. Sulfur (S) containing soil amendments are more often applied to lower the pH of orchard soils above 8.0. Sulfur deficiency has been observed in wheat and other cereals in the Sacramento Valley during late winter-early spring when rainfall is high for several weeks, soils become saturated, sulfur is chemically reduced, and the supply of SO4 2 -S becomes insufficient to meet crop needs.

As soon as the soils dry out and warm up for bacteria to convert the sulfur to the SO4 2 -S form, the sulfur deficiency disappears. Sulfur nutrient deficiency is uncommon in orchard crops grown in the Sacramento Valley and other areas of California. Still, S fertility levels are evaluated and reported in soil tests. Water soluble SO4 2- may be measured in the saturated paste extract using a barium chloride turbidimetric method or estimated by subtraction if electrical conductivity (ECe), chloride, bicarbonate, and carbonate have been measured in the saturation paste extract. Alternatively, SO4 2 -S may be measured and reported for a soil sample using a calcium phosphate extraction method. Both of these methods of soil testing have limits in predicting favourable crop responses to sulfur applications in orchards.

These methods are subject to the time of sampling because S availability changes in the soil due to mineralization and leaching and is complicated by the fact that several sources of S besides the soil itself (i.e. atmosphere, rain, and foliar applications such as fungicides) can contribute to the needs of trees. As a result, if S tree nutrition is in question, tissue testing should be used for a more positive diagnosis.

Sulfur: Sulfur is absorbed by the roots of plants in the form of SO42- ions and by the leaves in the form of SO2 gas.

Deficiency symptoms of Sulfur in plants

1) Sulfur enhances the resistance against diseases in plants.
2) Sulfur develops roots and increases the number of tillerings in crops.
3) Sulfur increases oil content in oilseed crops by 4-9%.
4) Sulfur is essential for the synthesis of essential oils in plants.
5) Adequate amount of sulfur in the field increases the yield of tobacco by 15-20%.
6) The formation of root nodules is accelerated in pulse crops.

7) Sulfur is essential for onion, garlic and mustard crops. The odor coming from onions and garlic is formed due to sulfur.
8) Sulfur helps in the formation of chlorophyll in the leaves of plants.
9) Sulfur produces a characteristic odor in plants of the Cruciferous species.
10) The presence of sulfur increases the absorption efficiency of nitrogen and phosphorus in plants.
11) Sulfur helps in the formation of vitamins biotin, thiamine and B1.
12) Sulfur is an essential component in the formation of glutathione.

Deficiency symptoms of sulfur in plants

1) Symptoms of sulfur deficiency in plants are similar to those of nitrogen, although the symptoms of sulfur deficiency first appear on the upper part of the plants because sulfur is not motile in plants.
2) Due to the deficiency of sulfur, the veins of the upper leaves (new leaves) of the plant and the parts between the veins become light greenish yellow.
3) The stems and leaves of the plant become short, thick and hard.
4) Flowers and fruits are less in banana, coconut etc.
5) Potato, mustard and tea leaves become bowl-like (cupping).
6) In cotton, lucerne, rapeseed, maize, tomato the stem or stalk becomes red in colour.
7) Root nodules are less formed in leguminous crops.
8) Sulfur deficiency occurs in areas where there are sandy lands and organic matter is less or rainfall is more. In irrigated areas and sandy lands, sulfur is lost in the form of sulphate due to leaching

Micronutrients and Beneficial Elements- Role, Dynamics and Management

Plant nutrients are essential for plant growth and development. According to current scientific knowledge, there are 14 essential plant nutrients, which are divided into macro- and micronutrients. Plants furthermore absorb other elements, which scientists call beneficial elements. Here is information on these groups and the important functions of potassium, magnesium, boron, manganese & Co.

Macronutrients High nutrition demand: Macronutrients are present in plant tissues in comparatively high concentrations and the requirement for healthy growth corresponds to several kg of plant nutrient/ha. There are six macronutrients: potassium, magnesium, sulfur, nitrogen, phosphorus, and calcium.

Micronutrients Small amount with great effect: Micronutrients occur in plant tissue in comparatively low concentrations. The plant requirement is several grams of plant nutrient per hectare. They are also referred to as trace elements. Today, eight micronutrients are known: boron, chlorine, iron, copper, manganese, molybdenum, nickel, and zinc. Here is information on the functions of the most relevant micronutrients:

Boron: Boron improves stability and function of cell membranes and supports plant growth by stimulating cell division. The requirement of the various crops for boron is very different.

Copper: As a component of several enzymes, copper has a positive effect on the plant metabolism. For instance, it regulates the photosynthetic electron transport.

Manganese: Manganese either stimulates, or is a component of many enzymes and, therefore, greatly affects the metabolism of the plant. It affects photosynthesis, synthesis of fatty acids and carbohydrate metabolism.

Zinc: Zinc activates or is a component of several enzymes and therefore affects many metabolic processes in the plant. For instance, as an essential component of RNA polymerase it affects RNA and protein synthesis.

Beneficial Elements

Helpful, But Replaceable

Unlike macro- and micronutrients, beneficial elements are not essential for plant growth and development. They can perform important functions in yield, quality, or stress tolerance, but can be replaced by other elements. Nevertheless, they often play an important and essential role in humans and animals living on plants. Examples are sodium, selenium, or silicon.

Sodium: Sodium increases fodder productivity and keeps animals healthy and fertile. Within the plant, sodium controls osmotic pressure in the cells which results in a more efficient use of water.

Nutrients and beneficial elements: When is a nutrient essential-

Plant nutrients - unlike beneficial elements - are essential for plant growth and development. Macro- and micronutrients meet the criteria applying to them:

- If plant nutrients are not available in sufficient quantities, this leads to a specific deficiency symptom.
- Each plant nutrient has at least one function within the plant that has been scientifically proven and for which it cannot be replaced by any other element.
- If a plant nutrient is completely lacking, growth comes to a standstill and the plant cannot complete its life cycle.

According to these criteria, there are 14 essential plant nutrients according to current scientific knowledge. In addition, plants take up other elements that are not essential for their growth and development and do not meet the three criteria for essential plant nutrients mentioned above. These substances are referred to as "benefical elements".

Figure: *Balanced fertilization of crops*

Figure: *Optimize your fertilization*

For all important crops, we provide you with the decisive recommendations for targeted nutrient supply - for the best quality and high yields.

Use our free interactive tools to quickly and easily optimize your fertilization: Identify and correct nutrient deficiencies with the 1x1 of deficiency symptoms, determine fertilizer requirements with the Liebig nutrient calculator, and calculate different forms of nutrients in fertilizers with the nutrient converter.

Organic Manures-Production and Enrichment

Agriculture intensification through indiscriminate and irrational se of chemical fertilizers and pesticides has resulted in reduction of crop productivity, lowered fertilizer use efficiency, accelerated environmental degradation and soil heath thus posing serious threat to the sustainability of agriculture, ecological balance and human health. However, people are gradually realizing the emerging danger and showing interest in sustainable crop production practices such increased use of organic manures to revitalize and restore soil fertility and reviving the microbial activity of the soil.

Traditionally farmyard manure, animal wastes, compost, vermicompost are used as source of organic fertilizers for improving soil fertility and crop productivity. However, low nutrient content, bulkiness, handling difficulties and labour intensive application prevent the growers for larger use of these traditional organic manures. Gradually there is a decline in microbial activity of these organic manures which in turn leads to depletion of enzymatic activity and composition of nutrient. Manure enrichment increases the nutrient content, microbial population, enzyme activity and increase the fertility status of the soil.

Introduction: Organic farming practices have gained significant popularity worldwide due to their numerous environmental and health benefits. Central to these practices is the use of organic manure, which plays a vital role in enhancing soil fertility and promoting sustainable agriculture. This blog discusses the meaning of organic manure, discusses its various types in India, and highlights its importance in modern farming practices.

What is Organic Manure: Organic manure, which is also known as organic fertiliser, refers to natural substances derived from plants, animals, or their by-products. It is composed of organic matter that decomposes over time and releases essential nutrients into the soil, making it more fertile. Unlike synthetic fertilisers, organic manure is free from harmful chemicals and provides a holistic approach to soil enrichment.

Types of Organic Manure in India: There are various types of organic manure used in India. Some of the common organic manure examples are farmyard manure, vermicompost, and green manure, among others. These are discussed below:

Farmyard Manure (FYM): Farmyard manure (FYM) holds significant importance as a widely utilized organic manure in India, and it is the most common organic manure examples. Its production involves the decomposition of a blend comprising animal dung, crop residues, and farm waste. This process generates a valuable resource that is abundant in organic matter, essential nutrients, and beneficial microorganisms.

It can also be considered as natural manure, and when applied to agricultural fields, FYM contributes to enhancing soil structure, water retention capacity, and nutrient-holding capacity. These improvements, in turn, result in increased crop productivity. By providing a balanced and natural source of nourishment to the soil, farmyard manure plays a crucial role in sustainable farming practices and the overall health of agricultural systems.

Vermicompost: Vermicompost is another organic manure example, which is a nutrient-rich organic manure that is created through the decomposition of organic waste materials by earthworms. This natural process results in a high-quality fertiliser that offers several benefits for soil fertility and plant growth. Vermicompost enriches the soil by providing essential nutrients and micronutrients in a readily available form.

It also enhances microbial activity in the soil, supporting beneficial bacteria and fungi that contribute to plant health. Vermicompost improves soil structure, promoting better aeration and water-holding capacity. It can help suppress certain plant diseases and pests, reducing the need for chemical interventions. ***Green Manure:*** Green manure is an agricultural practice that involves cultivating certain plants with the purpose of improving soil fertility. This can also fall under the natural manure category.

Leguminous plants such as clover, alfalfa, and soybean are commonly chosen for green manuring. These plants possess the ability to fix atmospheric nitrogen through symbiotic relationships with nitrogen-fixing bacteria, enriching the soil with this essential nutrient.

When green manure crops are ploughed back into the soil, they contribute organic matter, further enhancing soil structure and nutrient content. Green manure crops also act as natural weed suppressants, reducing the competition for resources. They also help prevent soil erosion by improving ground cover.

Bone Meal: Bone meal is an organic manure produced by grinding animal bones, and it serves as a valuable source of phosphorus and calcium for plants. Phosphorus is an essential nutrient for various aspects of plant growth, including root development, flower formation, and fruit production. Calcium, on the other hand, plays a crucial role in cell structure and nutrient transport within plants.

Bone meal gradually releases these nutrients into the soil, providing a long-lasting and consistent supply. This slow release ensures that plants have access to phosphorus and calcium over an extended period, promoting healthy growth and development.

Fish Emulsion: Fish emulsion is a liquid organic fertiliser that is derived from the fermentation of fish waste or fish remains. This process produces nutrient-rich manure that is abundant in nitrogen, phosphorus, and trace elements.

The high nitrogen content of fish emulsion promotes lush and vigorous plant growth, while phosphorus supports flower and fruit development. Fish emulsion is easily absorbed by plants, allowing for quick nutrient uptake and utilization. This results in healthy growth, vibrant flowering, and increased yields.

Importance of Organic Manure: With the increase in demand for organic foods and products, the organic farming in India is being increasingly adopted by farmers wherein the use of organic manure is must. In this section, we will see the importance of organic manure.

Improved Soil Fertility: One of the major importance of organic manure is that it enhances soil fertility by replenishing essential nutrients, improving soil structure, and promoting beneficial microbial activity. It aids in maintaining long-term soil health, preventing nutrient depletion, and ensuring sustainable agricultural practices.

Environmentally Friendly: Organic manure, being free from synthetic chemicals, plays a significant role in promoting a healthier environment. This is one of the advantages of organic manure. Its use minimizes the risk of soil and water pollution that can occur due to the excessive application of chemical fertilisers.

Unlike chemical fertilisers, organic manure is derived from natural sources such as animal waste, plant residues, and composted materials. These organic inputs are rich in nutrients and support the growth of beneficial microorganisms in the soil.

Sustainable Agriculture: One of the crucial importance of organic manure is that it indeed aligns with the principles of sustainable agriculture by reducing reliance on synthetic inputs. Its use promotes a balanced and natural approach to farming.

By utilizing organic manure, farmers can maintain soil productivity through the gradual release of nutrients and the improvement of soil structure.

This, in turn, reduces the need for synthetic fertilisers and helps preserve the long-term fertility of the land. Also, organic manure aids in conserving natural resources, such as water and energy, by minimizing the use of chemical-intensive practices.

Nutrient-rich Food: Organic manure plays a crucial role in facilitating the production of nutrient-rich crops. The gradual release of nutrients from organic manure ensures a steady and balanced supply, allowing plants to access nutrients as needed throughout their growth cycle.

This results in crops that are inherently rich in essential nutrients, vitamins, and minerals. As a result, the harvested produce tends to have higher nutritional content compared to crops grown with synthetic fertilisers.

In addition to the nutritional benefits, organic farming practices contribute to better taste, flavour, and overall quality of food. The natural and diverse nutrient profile provided by organic manure enhances the flavour and aromatic compounds in fruits, vegetables, and herbs.

Soil Conservation: Organic manure plays a critical role in soil conservation by enhancing various aspects of soil health. By incorporating organic manure, the water-holding capacity of the soil is significantly improved. The organic matter present in the manure acts as a sponge, effectively retaining moisture and reducing the risk of drought stress for plants.

Organic manure also acts as a protective layer, reducing erosion by preventing the detachment and transport of soil particles by wind or water. This preserves the integrity of the soil, preventing valuable topsoil from being lost.

Organic manure contributes to the improvement of soil structure, promoting better soil aggregation and porosity. This allows for improved root penetration, nutrient uptake, and aeration, fostering the development of healthy soil ecosystems.

Conclusion: Organic manure plays a pivotal role in organic farming practices in India. Sustainable agriculture, soil fertility improvement, and environmental well-being promotion are some of the common advantages of organic manure.

With the growing awareness of the harmful effects of synthetic fertilisers, organic manure serves as a viable alternative that supports the production of healthy, nutrient-rich food while safeguarding the long-term health of our ecosystems. Embracing organic manure as a key component of agricultural practices is crucial for building a more sustainable and resilient future.

Enrichment of manures and composts is done by incorporation of mineral additives (rock phosphate, zinc salt etc.), bio-fertilizers (nitrogen fixing, phosphorus and potassium solubilising bacteria etc.) and bio-inoculations (Trichoderma, Pseudomonas etc.) have lot of benefits on crops in terms of higher yield, better quality, increased disease and pest resistance, availability of essential nutrients and increased soil microbial population and activity. Regular use of higher amount of enriched organic manures will not only reduce the amount of organic manure requirements, but will also increase the use efficiency of the applied manures. Preparation and application of enriched organic manures should be promoted to more number of vegetable growers for organic cultivation as well as for traditional crop cultivation.

Enrichment with beneficial microorganisms may increase the benefits of organic compost. The aim of this study was to evaluate the enrichment of a mature compost with plant growth-promoting bacteria and arbuscular mycorrhizal fungus (Rhizophagus clarus), using brachiaria (Urochloa brizantha) as host plant, totaling seven treatments: control (compost with no bacteria, brachiaria or mycorrhizal fungus); compost + brachiaria; compost + brachiaria + mycorrhizal fungus; compost + brachiaria + mycorrhizal fungus + Azorhizobium sp.; compost + brachiaria + mycorrhizal fungus + Azoarcus sp.; compost + brachiaria + mycorrhizal fungus + Bacillus subtilis; and compost + brachiaria + mycorrhizal fungus + Azotobacter sp., in a completely randomized design with three replicates. Brachiaria shoot biomass, N and P concentrations, mycorrhizal colonization, and chemical characteristics of the compost were assessed five times over 183 days.

B. subtilis and Azotobacter increased ammonium-N concentration in the compost in two and three sampling dates, respectively. In contrast, Azotobacter and Azoarcus decreased the concentrations of nitrate-N in at least one sampling. Despite high P availability in the compost (951-2927 mg kg-1), mycorrhizal colonization reached up to 53%. In a field trial with the produced composts, in a randomized block design with six repetitions, the composts with brachiaria doubled the mycorrhizal colonization of corn (Zea mays) and wheat (Triticum aestivum), independent of the growth-promoting bacteria and, depending on the associated bacteria, increased grain yields.

Composting allows the biological stabilization of organic residues of different origins, which can be used as source of nutrients for crops (Andrade et al., 2018). In addition to the beneficial effects of organic compost in improving soil fertility (Heck et al., 2013; Mota et al., 2019), enrichment with beneficial organisms like plant growth-promoting microorganisms may further increase the agricultural and environmental benefits of organic compost (Sousa et al., 2018). Plant growth-promoting bacteria (PGPB) and arbuscular mycorrhizal fungi (AMF) may interact synergistically (Pereira et al., 2013) and promote plant growth by several mechanisms including P-solubilization, P-mineralization, biological nitrogen fixation (BNF), nutrient cycling, synthesis of siderophores, and phytohormones (Rodrigues et al., 2012). Moreover, some PGPB, also called "mycorrhiza-helper bacteria", may stimulate root mycorrhizal colonization (Choudhary et al., 2017). Some PGPB may perform free-living nitrogen fixation and enrich the compost with N (Sousa et al., 2018), whereas AMF may reach microsites with their external hyphae that are inaccessible to root hairs, thus increasing the uptake of low-mobility nutrients such as P (Brito et al., 2017). The tripartite interaction among the plant, AMF, and PGPB may result in a synergistic effect on plant performance (Pereira et al., 2013). Thus, enriching organic compost with beneficial microorganisms may yield even better results than the commonly used composting process.

During the era of Green Revolution, introduction of high-yielding varieties, extension of irrigated areas, use of high analysis NPK fertilizers and increase in cropping intensity, boosted the production in most of cases, propelled India towards self-sufficiency in food production. In the process, relative contribution of organic manures as a source of plant nutrients vis-a-vis chemical fertilizers declined substantially. One way of replenishing nutrients in the arable lands is to recycle nutrients through application of organic material such as litter, crop residues, and manures. Organic manures, especially farmyard manure, have a significant role for maintaining and improving the chemical, physical and biological properties of soils. Organic materials play a critical role in sustainability in the arid and semi-arid regions. Despite this importance, there is little predictive understanding for the management of organic inputs in arid and semi-arid agro ecosystems. It is now widely recognized that soil organic matter plays an important role in soil chemical (pH, base saturation, salinity and CEC changes) physical (bulk density, stabilization of soil structure and aggregate formation) properties and biological properties. Farm yard manure prepared from cattle manure is the most important organic soil amendment utilized in India. Organic amendments, such as FYM are known to improve soil physical properties. Organic matter is an important soil constituent influencing a number of constraints linked with crop productivity. The loss of soil fertility, in many developing countries, due to continuous nutrient depletion by crops without adequate replenishment poses an immediate threat to food and environmental security. Intensive cropping and tillage system have led to substantial decreases in soil organic matter levels of much prime land in the world.

This decrease in soil organic matter levels seems to be associated with the decline in soil productivity. Hence, the application of FYM is essential for maintaining soil fertility. Use of enriched FYM is one of the methods to decrease the nutrient losses (Aswini et al. 2015). The aim of enrichment of organic manures is to minimize excess use of fertilizers for optimum yield and quality of crops without harming soil and environment health by the application of PGPR and plant nutrient enriched FYM. ~ 926 ~ International Journal of Chemical Studies Farm Yard Manure The term FYM refers to the decomposed mixture of dung and urine of farm animals along with the litter and left over material from roughages or fodder fed to cattle. FYM collected from dairy / cattle shed consists mainly of dung and part of the urine soaked in the refuses. Table: Nutritional status of FYM (%) Nitrogen 0.5000 Phosphorus 0.2500 Potassium 0.4000 Calcium 0.0800 Sulfur 0.0200 Zinc 0.0040 Copper 0.0003 Manganese 0.0070 Iron 0.4500 Enriched FYM FYM is enriched by the addition of chemical preservative, which helpful to reduce the nitrogen losses. The preservatives are applied in the cattle shed to permit direct contact with the liquid portion of excreta or urine, as loss of N from urine starts immediately. The commonly used preservatives are Gypsum and Superphosphate. Ammonium carbonate is the immediate product of decomposition of urea in urine, which reacts with Gypsum. The resultant outcome of the reaction is ammonium carbonate is changed in to Ammonium sulphate which is a stable salt. The reaction is as under: (NH4)2 CO3 + CaSO4 CaCO3 + (NH4)2 SO4 Ammonium Carbonate Gypsum Calcium Carbonate Ammonium Sulphate Super phosphate has been extensively used as a manure preservative, since superphosphate contain 50 - 60 % Gypsum (Calcium Sulphate) besides monocalcium phosphate. The action of SSP is similar to that of Gypsum. In Indian conditions due to high temperature FYM becomes dry, the use of SSP could be safely recommended. Enrichment with micronutrients and PGPR The known quantity of FYM was filled in four pits. The solution of ZnSO4.7H2O was prepared on the basis of 2.5 and 5.0 kg Zn ha-1 (according to soil test) and thoroughly mixed in respective pits of known quantity of FYM as required of one hectare of soil. After 21 days, a composite microbial culture consisting of PGPR consortium (Azospirillum brasiliense, Pseudomonas fluorescens, Pseudomonas aeruginosa, Bacillus subtilis, Paenibacillus polymyxa, Trichoderma harzianum) was added and thoroughly mixed to ensure complete contact with the decomposing materials (FYM). PGPR consortium was added after 21 days of initial composting to protect the microbes from direct exposure to excess heat generated from the materials. The moisture percentage of organic manures after mixing with ZnSO4.7H2O and PGPR was maintained around 60% throughout enrichment process. The cattle dung slurry @ 1% was applied as starter inoculums of microorganism to boost up the microbiological activities for enhancement of natural process of composting to fix the externally added Zn through zinc sulphate. The pits were covered by polythene sheet and allowed to decompose. The mixture was turned over weekly and also maintained moisture (Regar and Yadav.

Effect of Enriched Organic Manure on Crop Growth and Yield

Debele et al., (2000) found that application of enriched FYM either 4 or 8 t ha-1 significantly improved the grain yield of maize by 32.6 and 22.4 per cent over the respective levels of conventional FYM. Basavaraj and Manjunthaiah (2003) found that the combined application of P-enriched organic manure at 100 percent RDF-P had significant effect on grain yield of maize, which accounted 27.50 per cent increase in yield over control. Field experiment was conducted during rabi season of 2009- 10 on an established plantation from 2006 at Agroforestry Research Centre, Haldi, G. B. Pant University of

Agriculture and Technology, Pantnagar to evaluate the suitable sources of phosphorus nutrition on the yield of understorey wheat crop by Chaturvedi et al., (2011) . Rock phosphate (RP) enriched with FYM + PSB had significantly highest wheat yield among all the treatments however, it was statistically at par with RP + PM + PSB. Rock phosphate (RP) enriched with FYM + PSB had recorded 16.49 % higher yield over SSP alone. Yield attributes like number of spikes, number grains per spike, 100 grain weight were found maximum with RP + FYM + PSB being statistically at par with all enriched treatments except RP alone and RP+ FYM.

Application of compost enriched with N and LTRP in the presence of 50% supplementary dose of N fertilizer significantly increased the fresh biomass (11.8%), cob weight (12.6%) and grain yield (13.6%) compared to full dose of N fertilizer. Similarly, application of enriched compost plus 50% N had significant increasing effect on root weight (15.7% more than the full dose of N fertilizer) of maize plants. In general, N plus L-TRP enriched compost gave better performance than the application of N enriched compost (Ahmad et al. 2007). At Anand (Gujarat) application of Zn and Fe enriched FYM improved average mustard seed yield by 20 per cent over control whereas it was 11 per cent over straight Zn and Fe application.

The Zn and Fe enriched FYM enhanced uptake of N, S and micronutrients by mustard, and improved oil and protein content of mustard seed. The yield of subsequent sorghum (fodder) increased by 11 per cent due to residual effect of FYM over control (Meena et al. 2008). A significant increase in grain and straw yield was noticed in the treatments which received Zn-E compost at 15 kg ha-1 (54.06 and 78.01 q ha-1) and 10 kg ha-1 (53.96 and 78.90 q ha1) it was on par with package of practice (55.80 and 79.17 q ha-1 by Veeranagappa et al., (2010). A field study conducted to find out the effect of zinc-enriched organic manures.

The treatments were NPK alone, NPK +200 kg FYM without enrichment, NPK + 200 kg FYM enriched with different levels of Zn (1.25 kg, 2.50 kg and 5.0 kg Zn ha1), NPK + cow dung without enrichment, NPK + 200 kg cow dung enriched with different levels of Zn. The result of the field experiment revealed that recommended dose of NPK +200 kg FYM enriched with 5.0 kg zinc increased the grain (5430 kg ha-1) and straw (7075 kg ha-1) yields which was due to increased availability of zinc in soil (Sridevi et al. 2010). A pot study was conducted in Institute of Soil and Environmental Sciences,

University of Agriculture, Faisalabad, Pakistan during 2010 to compare Zn enriched farm yard manure (FYM) with chelated (Zn-EDTA) and non chelated (zinc sulphate, ZnSO4) Zn sources for optimum grain yield and higher Zn concentration in grains of rice (cv. Shaheen Basmati) grown in a salt-affected soil. Treatments ~ 927 ~ International Journal of Chemical Studies comprised of 1.5, 3 or 6 mg Zn kg-1 soil in the form of both ZnSO4 and Zn-EDTA, and Zn-enriched farm yard manure 5% of soil (w/w). Zn enrichment was done with same three levels of Zn using ZnSO4.

Application of Zn increased plant growth, grain yield and Zn concentration in various tissues, irrespective of the Zn source. The increase in plant growth and Zn concentration was higher when plants were grown with Zn enriched FYM compared to ZnSO4 or Zn-EDTA application alone. The maximum Zn concentration in rice grains (13.9 mg kg-1) and straw (19.1 mg kg-1) were obtained with the same treatment. It was concluded that application of ZnSO4 enriched FYM proved better over application of alone ZnSO4 or Zn-EDTA indicating the positive role of organic matter in increasing grain yield and grain Zn concentration on soils affected with salts and depleted in organic matter (Ahmad et al. 2012). Mali et al., (2015) observed that the maximum plant height (206.73 cm), dry matter (164.67 g plant), average cob length (16.87 cm), no. of grains cob (391.00 grain cob), cob weight (134.67 g) and test weight (197.0 g) were recorded under application of 25% RDP through PRC + vermiculture + PSB + 75% RDP through DAP.

The maximum grain yield (3.93 t ha), stover yield (5.74 t ha), net return (38337 Rs ha) and B:C ratio (2.19) were recorded under application of 25% RDP through PRC + vermiculture + PSB + 75% RDP through DAP and lowest grain yield (1.89 t ha), stover yield (3.14 t ha), net return (14611 Rs ha) and B:C ratio (1.14) were noted in control where phosphorus rich compost was not applied. Maximum yield response (107.5%) was recorded with 25% RDP through PRC + vermiculture + PSB + 75% RDP through DAP followed by 25% RDP through PRC + PSB + 75% RDP through DAP (95.7%). A Field experiment to assess the effect of enriched farm yard manure (FYM) and fertilizer levels on growth and yield components of aerobic rice was carried out at university of Agricultural and Horticultural Sciences, Shivamogga during kharif, 2013 by Ashwani et al., (2016). Among the methods of application, spot application of enriched manure recorded significantly higher 1000 grain weight (24.24 g) and grain yield (54.03

q ha-1) due to timely available of nutrients. Interaction of spot application of enriched manure with 125:62.5:62.5 kg NPK ha-1 has registered higher grain yield (60.58 q ha-1) and filled grains panicle-1 (128.75). While, higher 1000 grain weight (25.80 g) was recorded in spot application of enriched manure with 100:50:50 NPK kg ha-1 . An experiment was conducted in three rabi seasons from 2010-11 to 2012-13 at Center for Research on Seed Spices, S. D.

Agricultural University, Jagudan to find out the effect of Fe and Zn enriched with farm yard manure on growth and yield, quality and uptake of nutrients for cumin. Enrichment of Fe or Zn or both with FYM (T8, T7 and T6) increased not only seed yield but net income and BCR also. Combine application of Fe and Zn found better than alone application of Fe or Zn. Moreover, enrichment of Fe and Zn with FYM not only reduced the 50 per cent requirement of micronutrient but increased the yield (577 kg ha-1) considerably also.

An application of RDF along with 1.0 t FYM enriched with 1.5 kg Fe and 0.75 kg Zn ha-1 (T8) registered significantly higher uptake of Fe (64.70 g ha-1) and Zn (24.60 g ha-1) by seed (Patel et al. 2016). Regar and Yadav (2017) found that application of PGPR and Zn @ 5.0 kg ha-1 enriched FYM along with 60 kg P2O5 ha-1 resulted in markedly higher plant height (109.07 cm), grain yield (59.62 q ha-1), straw yield (79.94 q ha-1) and biological yield (139.56 q ha-1) but showed at par result with PGPR and Zn @ 2.5 kg ha-1 enriched FYM applied in conjunction with 40 kg P2O5 ha -1. The aim of this experiment was to minimize excess use of fertilizers for optimum yield and quality of rice without harming soil and environment health by the application of PGPR and Zinc enriched FYM.

Effect of Enriched Organic Manure on Soil properties Improvement in the nutrient uptake with the use of enriched FYM either alone or in conjunction with fertilizer was observed by Debele et al., (2000). Soil organic carbon declined while there was marginal increase in available N due to enriched FYM. Kumar et al., (2004) found that incorporation of farmyard manure at 12.5 t ha-1 along with Zn solubilizing bacteria stood superior by registering highest values for available of N, P and K content in the soil.

The DTPA - Zn content of the soil though evidenced significant variation for the application of FYM, FYM + ZSB, Zn + Fe on an overall basis the actual values did not exceed the deficiency level. Ahmad et al., (2007)reported significant improvement in N, P and K contents of maize plant in response to application of N and L-TRP enriched compost supplemented with 50% N fertilizer. Maximum increases (10.1% N, 11.9% P and 7.5% K) were found in response to combined application of enriched compost and N fertilizer compared to sole application of full dose of N fertilizer.

The effect of N enriched (without L-TRP) compost plus 50% N was statistically similar to that observed in case of full dose of N fertilizer alone. Nitrogen, phosphorus, potassium and Zinc uptake significantly differed among the treatments. Recommended NPK + 200 kg FYM enriched either with 5.0 kg Zn ha-1 or 2.5 kg Zn ha-1 was superior in recording higher nitrogen uptake than control, but on par with each other treatments and showed statistical parlance among themselves (Sridevi et al. 2010). Different rock phosphate enriched treatments had nonsignificant effect on soil properties after cop harvesting, however, Organic carbon, Available N, P and K were found to be higher under tree as compared to control. The tree mean diameter (cm tree-1) at standard (1.37m) breast height was maximum under RP + FYM followed by SSP alone.

However, Height was highest in RP+FYM+PSB followed by RP + Pressmud (Chaturvedi et al. 2011). A slight improvement in soil pH, electrical conductivity and organic carbon content noticed higher values in NPK + zinc enriched levels followed by package of practice. The application of Zn-E compost increased the amount of zinc present in different fractions. Increase in amount of zinc present in different zinc fractions might be due to the higher solubility and mobility of the added zinc source. Water soluble plus exchangeable fractions significantly increased with the enrichment of compost (Veeranagappa et al. 2011). The availability of all the three major nutrients has conspicuously improved for the different treatments.

Incorporation of farmyard manure@12.5 t ha-1 along with Zn solubilizing bacteria stood 84 Effect of Zn enriched organic manures and zinc solubilizer application on yield, curcumin content of turmeric superior to FYM alone and control treatment by registering the highest values for availability of N, P and K content in the soil. As compared to control, per se addition of FYM also exerted positive effect on the major nutrients availability, which was quite natural since it has been well established that bulky organic manures play a vital 928 International Journal of Chemical Studies role in improving the major nutrient availability by direct contribution as well as indirectly by influencing chemical transformation reactions and microbial activity.

The Zn solubilizing organism (Bacillus sp.) proved to have a favourable effect on the availability of N, P and K, thereby indicating the vital role of these organisms in the transformation reaction of these three nutrients in the soil.

The DTPA - Zn content of the soil though evidenced significant variation for the application of FYM, FYM + ZSB, Zn + Fe on an overall basis the actual values did not exceed the deficiency level. However at 90 DAS and 120 DAS the treatments receiving ZnSO4 as such and also as Zn fortified FYM / coirpith reached the level of sufficiency (Regar and Yadav, 2017).

Chapter 6

Green Manures and Crop Residues: Production and Management

Introduction

The global demand for food will grow considerably in the coming years due to the increasing global population that is supposed to reach 9-billion people by 2050. The agriculture practiced in the tropics has key importance on food supply for much of the current global population and may become even more important for future generations.

The soils in most of these tropical environments have high acidity and aluminum toxicity, and are rich in oxides and poor in nutrients. Therefore, the use of lime and fertilizers accounts for a large part of the agricultural production cost. Thus, to increase the environmental and economical sustainability of these environments, it is important to make rational use of fertilizers and find viable alternatives to maintain a good physical, chemical, and biological soil characteristics.

We highlight the use of green manures and crop residues as practices that can help maintain or increase the productivity capacity of the soils, since they act as conditioners of the physical, chemical, and biological characteristics. Since the ancient Greeks, Romans, and Chinese, the humanity wisely used fresh organic matter as green manures in order to maintain the land productivity, and even today, this practice has been used with the same purpose. Meanwhile, crop residues were "a problem" for many years in agriculture. The removal and/or burning of residues were common practiced used in order to accelerate its degradation in conventional tillage. The expansion of no-till system, from the 1970s, presented several benefits by conserving and managing residues of annual crops, especially at medium and long term. The cultivation of perennial crops also presented major recent changes, such as management of weeds between plants, not using fire in renewing and especially the use of processed residues.

In this chapter, we present some of the benefits of management of green manure and crop residues, mainly the nutrient supply potential for crops of economic interest. Initially, we discuss the factors that most influence the fresh organic matter decomposition and nutrient mineralization in tropical areas. Subsequently, we discuss the concept of green manure, its management, the amount of nutrients potentially accumulated and their advantages and disadvantages. In the last part, we present examples from the main crops grown in the tropics. Regarding the annual crops, we focused on legumes with greater economic impact, since they are the first crops to be planted in rotation or succession managements. Regarding perennial crops, we present the contribution of the main crop residues and the processing of sugarcane, coffee, and eucalyptus.

Green Manures

Green manure can be defined as the practice of plant cultivation until the flowering stage or until the incomplete development of seeds, with subsequent cutting and/or incorporation of its biomass into soil. The basic purpose of this technique is to improve chemical, physical, and biological soil characteristics in order to increase or stabilize the production of one or more crops in an area. Famers usually use legumes as green manure because of the high biomass yield, biological nitrogen fixation (BNF), and cycling of nutrients from deeper soil layers.

After Green Revolution, the practice has gained importance, especially in organic production systems. However, the species used as green manure are not restricted to these systems. They are also used, for example, to control soil degradation in minimum tillage, no-till (NT) and in integrative systems. In some situations, these species are called "cover crops," since the main purpose may be to conserve the soil. Species from the Poaceae family are the most common used cover crops. A better soil cover and diversification of species in an area may also be used to suppress the development of weeds and also the emergence of pests and diseases. From an environmental stand point, the use of these plants can be a form of sequestering CO2 from the atmosphere. This practice can also improve the use of mineral fertilizers, especially for those nutrients more susceptible to losses by leaching or sorption.

The main species used as green manure, their benefits, limitations, and managements are discussed in the following topics. More information, you can find in the references.

Plants Used as Green Manure

Tropical environments allow the cultivation of a wide variety of green manure species, with plants from the families Fabaceae, Brassicaceae, Asteraceae, and Poaceae, among others. Some of the main species used in tropical environments, their characteristics of dry matter yield, amount of fixed nitrogen, and mineral composition of macronutrient and micronutrients is presented in Table 1.

Dry matter yield, amount of fixed nitrogen, and mineral composition of macro- and micronutrients in the shoot of species frequently used as green manure and cover crops in Southeastern Brazil.

Source: Wutke E.B., Trani P.E., Ambrosano E.J., Drugowich M.I. Adubação verde no Estado de São Paulo. Campinas: Coordenadoria de Assistência Técnica Integral-CATI; 2009. 89 p.

The first thing to consider in the process of choosing a green manure is the purpose of the farmer in using it. When the farmer objective is nutrient supply, plants with high biomass production and capable of associating with N-fixing bacteria can be suggested (Table 1). Among these plants are species from the genus Crotalaria, Canavalia, Cajanus, Leucaena, and Acacia. We recommend inoculation of seeds with Rhizobium in order to increase the symbiosis efficiency and consequently the biological nitrogen fixation. On the other hand, when the farmer objective is erosion control, species with higher C/N ratio, fasciculate root system, greater biomass production and lower shoot/root ratio can be suggested, such as some species of genus Pennisetum and Brachiaria. (Table 1).

Another important point to consider for choosing the correct green manures is the phytosanitary aspect. Some species can control plant pathogens, such as species from the Crotalaria genus, which are effective in controlling certain nematode species that cause root-knot . However, others are hosts of pathogen species and pests of commercial crops. An example of this problem is the cultivation of common beans (Phaseolus vulgaris) in succession or rotation with jack beans (Canavalia ensiformis). This green manure hosts the whitefly (Bemisia tabaci), an insect that transmits the bean golden mosaic virus. Therefore, the use of species from different phylogenetic groups is recommended. Moreover, the use of species from different families is convenient, since they have different patterns of nutrient accumulation and root architecture, enabling the exploitation of different soil layers.

Many green manure species are better adapted to the climatic conditions of spring and summer while others to the autumn and winter (Table 1). Sunnhep (Crotalaria sp.) and pigeonpea (Cajanus cajan), for example, have greater biomass and nutrient accumulation when grown between the spring and summer due to their water requirements, photoperiod, and/or thermoperiod. In general, sunnhep and pigeonpea plants require short days to flowering, so they have low biomass production when sown in late summer.

These plants should be selected for pre-cultivation management of commercial crop. Other species are more versatile, like the millet (Pennisetum glaucum), canavalia (Canavalia brasiliensis), and jack bean (Canavalia ensiformis), which have little or no sensitivity to photoperiod and good drought tolerance. Such species may have satisfactory dry matter yield in semi-arid conditions; however, they have to be sown, preferably, when there is still moisture in the soil for germination, such as the period after the summer harvest. On the other hand, the genres Avena, Lollium, Lathyrus, and Vicia have to be cultivated preferably between the autumn and winter, and some of them are better adapted to subtropical conditions. Therefore, the success in using the green manures technic depends on its adaptability to the region soil, climate conditions, cultivation system, and on their phytosanitary aspects. Experiments determining which species have greater biomass and nutrient accumulation are recommended when some of this information is unavailable.

Management of Green Manures in Tropical Conditions

Green manure plants can be used in minimum tillage, no-till or in systems with plowing and harrowing. These plants can be managed as soil cover in fallow, intercropping with perennial crops or in rotation or succession plans with annual crops. In general, when the plants reach the reproductive stage, when approximately 50% of the flowers are opened, a cut must be carried out with or without the plant matter incorporation into the soil. Some species might have longer cycles, but have to be cut before the seeds become viable. The black oat (Avena strigosa), for example, should be cut when grain maturation reaches the milky grain stage. Cutting these plants before this stage allows a regrowth, and after this stage the grains may become viable. Both cases may result in economic losses due to competition between the black oat plants and the main crop.

The toppling of plants in conservation systems can be obtained using mechanical or manual mowing, a roll-knife or herbicides for desiccation. In conventional systems, set of harrows are commonly used to incorporate the green manure into the soil or a rotary hoe is used in some cases. Plants from the Fabaceae family with C/N ratio around 20 (Table 2) are rapidly decomposed in tropical conditions. Therefore, the commercial crop planting in rotation/succession to this type of plant should be performed few days (less than two weeks) after the green manure tipping over and/or incorporation, since much of the N is mineralized in the first 60 days. Adopting species with higher C/N rates, around or above 40, such as species from the Poaceae family, enable to wait longer to the commercial crop planting. A 2-week interval between the grass cutting and the commercial crop planting is recommended, because of the low availability of N and possible allelopathic effects. Regarding the interim crops (Figure 1), the use of green manure species with climber habits should be avoided, since they may use commercial species as tutors. A solution to this problem is the use of more than one green manure species, a technique known as cocktail, using species with climbing and erect growth habit, so that the latter will serve as tutor for the first. It is also important an adequate inter-line plant density, in order to avoid competition for growth resources between the green manure and the commercial crop.

Figure: *Coffee seedlings with pigeonpea tipped over between planting lines. Photos by Lucas de Ávila-Silva.*

A study carried out in Brazil evaluated different green manures in rotation with maize under two managements, minimum or conventional tillage. The green manures used were as follows:

Crotalaria juncea, Canavalia ensiformes, Rafhanus sativus, mixed species (cocktail) with Crotalaria juncea, Canavalia ensiformes and Rafhanus sativus, and weeds (control). The green manures treatments had maize yield higher than the control with weeds, with average of 2.8-t/ha of peeled green ears. Another study, in the same region, evaluated the effect of N topdressing absence, pre-cultivation of C. juncea and topdressing with ammonium sulfate or urea on maize yield produced for silage. The use of C. juncea incorporated 12-days before the maize planting resulted in a yield increase of 14-t/ha compared with the control treatment without nitrogen topdressing. However, the treatment with green manure did not surpass the treatments with ammonium sulfate or urea topdressing, which increased silage productivity by 33 and 27-t/ha compared to the control without topdressing, respectively.

Therefore, in order to maintain a competitive agricultural system, the green manure practice must be managed together with chemical fertilizers. Amado et al., for example, proposed to consider the green manure contribution to define the nitrogen fertilizer dose, which is especially important in no-till and integrative systems. These authors considered the following criteria for setting the dose: the soil organic matter content; the commercial crop expected yield; and the previous cover crop (green manure) residue nature and amount.The previous cover crop contribution was considered in three situations: legumes, grasses in monocrop, and consortia (cocktails).They considered that the higher the green manure dry matter yield, higher the N supply to the commercial crop in succession to legumes. The contribution of grasses for the N supply was estimated to be very small or non-existent compared with uncultivated areas and may even reduce the availability of N due to its short term immobilization. Regarding the consortia system (cocktails), the contribution is estimated according to the percentage of legume biomass in total biomass of green manure. In some conditions, fertilizer savings can reach up to 80-kg/ha of N.

An example of the green manure potential intercropped with perennials plants is a study with soursop (Annona muricata) in southeastern Brazil, in which the author evaluated three green manures species (Gliricidia sepium, Crotalaria juncea, and Cajanus cajan). The greater amount of biologically fixed nitrogen was found with G. sepium (80% of the accumulated N) and with C. juncea (64.5% of the accumulated N). During the 2-years of experiment, the C. juncea added to the soil, in two cuts, about 149.5-kg/ha of N, from which 96.5-kg/ha came from biological nitrogen fixation (BNF). The G. sepium, managed with three annual pruning, added 113-160-kg/ha of N, with 90-128-kg/ha derived from BNF. Variations in natural 15N amounts indicated that green manure with C. juncea and G. sepium contributed to N supply for the soursop, transferring about 22.5 and 40% of the fixed N, respectively. They conclude that the use of these two green manure legumes contribute as organic fertilizer, supplying nutrients, mainly N.

The use of fresh organic matter from green manures brings more than just nutrients, affecting physical-chemical and biological characteristics of the soil. Pegoraro et al. showed the effect of the use of green manure (Acacia mangium) grown after the Eucalyptus cut in short rotation system (6-7-years of growth). The authors noted that the use of a legume in succession enabled the increased in stocks of total C and N, C and N in humic substances and in microorganisms compared to crops without the legume succession. Vegetable residues with lower C/N ratio accelerate the recycling of residues from commercial crops (ex.: Eucalyptus), which reduces the microbial attack on soil organic matter (SOM) and increases the levels of stabilized organic matter.

The effect on the soil organic matter supply makes the green manure a promising practice for recovery degraded areas and help to restore the A horizon of "beheaded" soil profile. Alves et al. studied green manures as a component of strategies for recovery degraded areas. Positive effects were noted 365 days after the experiment implementation, such as soil compaction reduction and water infiltration time reduction. Kitamura et al. reported, in another paper concerning the same research, that the treatments also provided results related to the soil macroorganism population.

Advantages and Disadvantages of Green Manures

The main advantages and disadvantages of green manures to tropical weathered soils are listed below.

Advantages

Chemical Aspects

1. Nitrogen input to the soil because of the green manure association with nitrogen-fixing bacteria;
2. Green manures with deep root systems allow cycling of nutrients that have been leached to deep layers;

3. Increased cation exchange capacity (CEC) due to an increase in soil organic matter content;
4. Green manures make possible to increase or stabilize the content *of soil organic matter;*
5. The release of organic acids allows the solubilization of more stable forms of phosphorus.

Biological Aspects

1. They favour the microflora and macroflora and fauna through carbon supply;
2. Some species control nematodes population;
3. They can serve for attracting insect pests and stop disease cycles;
4. They release compounds with allelopathic effect on weeds;
5. They compete for growth resources with weeds.

Physical Aspects

1. They promote protection against erosion by covering the soil;
2. They enhance stability of aggregates and porosity by adding organic matter and growth and death of roots;
3. They increase water retention by cover the soil and by add organic matter;
4. They allow natural decompression of the soil, when using species with deep root system;
5. They reduce the thermal soil amplitude.

Disadvantages

1. Inadequacy of some green manure species to the production system or the soil and weather conditions;
2. Lack of interest from consultants and farmers in this technology, which adopt immediate postures;
3. Sometimes, green manure involves costs with no direct financial return;
4. Low development of breeding technologies of green manure species;
5. Some green manures can host diseases and pests that attack the commercial crop;
6. Possibility of negative allelopathic effect of green manure residues on the commercial crop;

7. Possibility of competition between green manure plants and the commercial crop by inadequate management of the technology in intercropping systems;
8. Some green manures have incompatible decomposition rates with the nutrient requirements of crops;
9. Uneven seed germination of some species of green manure;
10. Difficulty of obtaining seeds for sowing;
11. Lack of functional decomposition models to predict nutrient release.

Crop Residues

Crop residues can be defined as any part of the plant without direct economic value produced in the field (derived from the harvest) or after processing on the farm. The crop residues value in agriculture coincided with the success of no-till system, until then, soil revolving or even fire was commonly used to accelerate residue decomposition. More recently, the crop-livestock-forestry integrated system is also an example of crop residues conservation. The no-tillage system implementation is based on three pillars: (i) no soil tillage, (ii) crop rotation, and (iii) permanent soil cover. Although essential, the soil cover maintenance in tropical environment is not a simple task, mainly due to the high rate of straw decomposition from leguminous plants, and is difficult to grow plants in succession (second crop) because adversities such as the rainfall patterns.

The maintenance of residues benefits the agricultural system by conserving the soil free from erosion. The homogeneous distribution of residues in the area with great amount of straw generates a slower runoff and less water and soil loss, consequently avoiding topsoil losses, which normally has a great amount of nutrients. Another important feature of crop residues is the maintenance of higher water content in the soil, a problematic factor in the tropical agriculture.

The crop residue maintenance also contributes to a decrease in the soil surface temperature, ensuring a better condition for the plants and other soil organisms. Another relevant factor is the ratio between the straw production, weed control, and soil compaction reduction capabilities, which is a recurring problem in poorly managed conservation systems.

The longer the time the crop residues are in the soil surface, the greater the positive effects. The residue permanence time in the soil depends on several factors, such as the: (i) fragmentation level, (ii) amount, (iii) chemical composition, (iv) contact level with the soil, (v) weather conditions, (vi) microbial community, and (vii) soil type. In addition to the factors discussed in the first part, it is also known that residues with greater contact surface with the soil accelerate the decomposition rate.

Green manures residues can be choosing from plants with high mass production, chemical composition with high C/N ratio and lignin, favouring the permanence of straw, the use of equipment which allows less fragmentation, such as the roll-knife.

Weeds and green manures (or cover plants) can be managed between rows of perennial crops; in this case, we can also use the roll-knife, although normally brush cutters and grinders are used.

Heavy equipment may be required in some situations, such as the scarifier or harrow, for example, in excessive amount of plant residues, rigid materials, or seeder inefficiency.

The residues management options for residues from commercial crops are less flexible, since the fragmentation is performed according to the harvest type, the amount is dependent on the yield and the material chemical composition is usually consequence of the best-adapted material. The legume residues have high decomposition rates, often cooperating little to the soil cover. The cultivation of legumes is usually performed first to leave a N balance for the sequential culture.

In specific cases of succession after grasses cultivation, an increase in N doses or even an anticipate application of nitrogen fertilizer is suggested before sowing, basically due to the N immobilization because the high C/N ratio of grasses.

After reminding here, the crucial importance of crop residues to the soil cover and the main factors affecting its decomposition, we can discuss a second advantage of residues from commercial crops: the potential supply of nutrients.

The management of residues from a preceding crop (annual) or residues from the crop itself (perennial) can complement the nutrient supply.

Regarding the green manure, nutrient mineralization from the crop residues depends on the residue quality, soil moisture and temperature, as well as specific soil factors such as texture, mineralogy and acidity, biological activity, and the presence of other nutrients. Most of the nutrients are exported in the harvest, remaining just a portion in the residues. In this part, we are going to see yield increases by the nutrient supply potential of the residues from the main crops in tropical environments.

Annual Crops Residues

The nutrient net mineralization from the annual crop residues that precede other crop must be considered in nutritional management. We focus here on annual crops that associate with N2 fixing bacteria, because they are the major contributor to the nutrient supply (nitrogen) for crops in succession or rotation. Subsequently, we will see the most used crops in tropical environments that leave nutrient to the system.

Soybean

The soybean (Glycine max) represents 50% of the global area of leguminous crops and 68% of global production of this family. The annual input of N fixed is 16.4-Tg, which represents 77% of N fixed by legume crops. A crucial fact to these crop success was the priority given to association with Bradyrhizobium and BNF in breeding programs.

Most of the soybean crop in Brazil is carried out in no-tillage system, and this management provides more root nodulation by the bacteria in the soil, best nodulation (nodulation deep in the soil profile), higher rates of BNF and yields compared to conventional tillage. These factors, combined with the efficient association with Bradyrhizobium, allow the non-use of nitrogen fertilizers in this crop.

Approximately 80-83-kg of N is required for each 1000-kg of soybeans produced, from which 51-65-kg are allocated in the seeds and 15-32-kg in the roots, stems, and leaves. The N2 fixation potential of the soybean can be as high as 360-450-kg/ha.

In southern Brazil, Paraguay, Uruguay, and northern Argentina (subtropical) is recurrent to get higher wheat yields (winter planting) when soybean is the preceding crop in the summer compared to the maize, precisely because of the remainder N. The maize, in rotation with soybeans, may have its nitrogen fertilizer rate reduced by 20% in this environment, considering the effect of rotation after a soybean crop with adequate productivity.

In a similar environment, some producers in various states of the Midwestern United States reduce 45 kg/ha of the nitrogen fertilizer dose, when maize is planted following soybeans compared to sequential crops of maize.

The Cerrado biome (Brazilian savanna) has no winter crops due to insufficient rainfall, and thus the soybean is the main summer crop and, after the harvest, in late summer, usually maize or sorghum is planted ("second crop"). A contribution of 20-kg/ha of N is assumed as a practical parameter in the region for these crops after soybean.

The first part of this chapter showed that the C/N ratio and biochemical constitution of the residues are important indicators to predict the nutrient mineralization rate. Regarding the soybean grown in tropical environments, different C/N ratios are observed in the roots (31.6), stems (22.5) and leaves (10.7), as well as larger amounts of lignin-suberin. These data explain the great decomposition rates of shoot residues in the first 20-days after harvest, with N mineralization, especially from leaf residues, under controlled conditions.

Unfortunately, there are few long-term field experiments in typically tropical regions, which hinder an accurate estimate of the decomposition rate of soybean residues and mineralization of nutrients. A field experiment carried out for 12-years in Brazil (Parana State, Brazil), estimated the residue decomposition percentage according to the time and management, with y-=-93.819e-0.0031x (R2-=-0.91) to the no-till and y-=-90.061e-0.0054x (R2-=-0.92) to the conventional tillage. The residue decomposition time in that work was autumn-winter, a less rainy period that the summer. Moreover, that is a climate transition region (Cfa subtropical climate in the Köppen classification) with low temperatures and a distinct pluviometric regime compared to Brazilian Midwest (Figure 2).

Figure: *Residue deposition and homogeneous distribution after the harvest. Photos by Lucas de Ávila-Silva.*

Beans

The genus Phaseolus has more than 200 species described, but those that have the greatest economic impact are the Phaseolus vulgaris, Phaseolus coccineus, Phaseolus lunatus, Phaseolus acutifolius, and Phaseolus semierectus, mainly the first. Different from the soybean, common beans have low symbiotic efficiency.

The nitrogen fertilizers use is indicated in some situations to achieve good yields, even though they may affect negativity the BNF efficiency. Mean values of 35% and maximum of 70% of N derived from the atmosphere were observed at the plant biomass considering six field experiments in tropical countries and Austria. Some progress is occurring with the best selection of strains, adapted to the conditions of each site.

Different C/N ratios were observed in stem fractions (79), straw pods (66), and senescent leaves (24) in four varieties of beans. A shorter leaf half-life was observed, although the straw pods also showed low half-life value (both of approximately 70-days). Those authors observed N and P release and half-life following the residue decomposition rate, while the K release was faster (average half-life of 18-days) and showed low difference regarding the material quality. In this case, N and P could be better used by the subsequent crop, while K can be reused by the same crop. The yield of 1350-kg/ha presented a potential for cycling 31.5-kg/ha N and 2.37-kg/ha P from the bean residues.

The species Vigna mungo, Vigna radiata, and Vigna unguiculata stand out in the Vigna genus, the latter being the most cultivated. These species have short cycle, low water requirement, and good development in low fertility soils; in addition, the BNF is capable of supplying more than 100-kg/ha of N. An increase in millet productivity of 9-24% was observed in an experiment carried out in three different locations in Niger, when cowpea was previous cultivated compared to successive cultivation of millet.

A C/N ratio of 15.8 was found in cowpea residues grown in a field experiment evaluating N mineralization depending on the phosphorus content in tissues. In the same experiment, the authors found an increasing in N mineralization (25, 32, and 34%) with increasing P concentration in tissues (1.0, 1.2, and 2-g/kg, respectively) in 8-weeks. This fact demonstrates the close relationship between nutrients interfering in the mineralization process, emphasizing the importance of fertilization management together with residues management. A potential mineralization from 6.8 to 9.2-g of N per kilogram of dry matter of cowpea residues is considered as a practical reference.

Groundnut (peanut)

India is largest producer of peanuts (Arachis hypogaea), which usually uses 10-20-kg/ha of N from ammonium sulfate. A large number of farmers in that country use crop rotation with groundnut to take advantage of its ability to improve soil fertility and increase the subsequent crop yield, due to the BNF.

The peanut cultivation in Brazil is also carried out with the same goal, but most in rotation with sugarcane. A C/N ratio of 15 and 24, and an addition of 70 and 38% of N from BNF in plant tissues, was, respectively, observed in the IAC-Caiapó and IAC-Tatu varieties, in acid soils. These authors emphasized that the N values from BNF for IAC-Tatu were low because the sampling, which was performed 120-days after sowing, when the plant was in an advanced stage of pods maturation and much of the N had translocated to the grains. About 90% of N was from the shoot in the average for these cultivars. In this same experiment, there was a yield increase of 12.2 (IAC-Caiapó) and 15.5-t/ha (IAC-Tatu) in the sugarcane crop compared with the control. Bagayoko et al. also reported an increase of approximately 39% in average yield of sorghum after groundnut cultivation compared to subsequent sorghum crops; during 3-years of experiment in Kouaré (Burkina Faso), they found N mineral increment available for sorghum.

The groundnut may also be infected by arbuscular mycorrhizal fungi, which favours the next culture infection and improves the P cycling.

Chickpea

The chickpea (Cicer arietinum) is the second more cultivated legume in the world to obtain grains. Most of the chickpeas production and consumption occur in developing countries (95%). A BNF evaluation by the 15N natural abundance method in different areas of Punjab showed that 58-86% of N (an average of 78%) could be derived from symbiosis, fixing 87-186-kg/ha of N. The average balance of N left by chickpea was 28-kg/ha N in that experiment, the yield ranged from 0.6 to 2.0-t/ha and the N in the soil increased in 38%, on average. The yield increase in the wheat planted in the sequence ranged from 19 to 73%, even with shoot residues removed. Around 90-kg/ha of N fixed was found in a cultivation field in Australia using similar methods.

Turpin et al. suggest that chickpeas can fix 146-214-kg/ha of N and contribute with 80-135-kg/ha of N, including the roots.

The ICRISAT work in Kenya reported that chickpea residues contributed from 30 to 35-kg/ha of N to the subsequent wheat crop, while the TSBF reported contributions of about 40-kg/ha of N for maize.

Difficulties

The absorption and mineralization must be synchronized for an effective N supply by a leguminous species. According Palm, organic residues release about 80% of their nutrients during decomposition, but <20% is absorbed by the crops, often because the lack of synchronization between the release and absorption. There are several studies reporting the decomposition rate and mineralization depending on soil and climatic characteristics, but there is still a lack of field data in some major producing regions and data to create models to predict the nutrients mineralization and availability.

The contribution of N mineralization arising from the root system is noteworthy, for example, Evans et al. and Larson et al. reported negative contribution of N to some tropical legumes when considering only the N from the shoot, with different results found when they accounted the N contained in the nodulated roots. Another factors to consider in the experiments are the biochemical characteristics and C/N ratios of the residues. We must pay attention to the fact that how higher residue decomposition rate, lower will be the vegetation cover time.

Perennial Crop Residues

Different from annual crops, nutrients from perennial crop residues are normally used in the same culture, but similarly, they take advantage of nutrients not exported at harvest. A perennial crops peculiarity is the use of residues from processing in the farm. Once seen as an environmental impasse, such residues are currently good sources of nutrients.

Sugarcane

The estimated BNF in sugarcane may reach 0.5-Tg per year. Although the N added by BNF generate savings for the system, the straw C/N ratio is around 97-149 and the stem around 118, promoting an initial immobilization of N. Straw is the main residue left in field, which has high C/N ratio and high lignin content, about 21%. It is estimated that each ton of sugarcane produces 150-kg of sugar, 140-kg of dry bagasse and 140-kg dry matter of straw, approximately 17-t/ha of straw.

In an experiment with 15N labeled straw, Gava et al. found that the N use efficiency by ratoon cane was 9% (68-kg/ha N) and that the main contribution of N from straw was to maintain or increase the organic N in the soil. Furthermore, this N became available in the second half of the cycle.

Another effective contribution of the sector is the vinasse return to the crops. Each liter of ethanol produced in distillery produces between 12 and 14-L of vinasse, which has 0.5-1.0% of soluble carbon and high levels of K (typically 12-g/L), and contain considerable amounts of other nutrients. Gava et al. used vinasse with the following contents (kg/m3): 0.41 of N, 0.07 of P2O5, 2.72 of K2O, 0.91 of CaO, 0.38 of MgO, and pH (water) of 4.9. Typically, the dose applied is around 80-100-m3/ha.

Resende et al. observed a sugarcane yield increase of 12-13% after application of vinasse in a period of 16-years. However, in some situations, vinasse is recurrently applied near of the distillery due to the cost of transportation, occurring concentrated applications with ground water salinization and/or contamination potential.

The filter cake is also a residue from the sugarcane processing used in the field. The sugarcane processing in Thailand produces 3.4% of filter cake and 25-30% of bagasse from the sugarcane fresh material. Each ton of sugarcane crushed generates around 40-kg of filter cake, which has variable composition, but with high levels of organic matter (OM), P, N, Ca, substantial amounts of K and Mg, as well as Fe, Mn, Zn, and Cu.

Two works illustrate the variability of nutrient content in the filter cake. In Brazil, Fravet et al. found pH of 4.5, C/N ratio of 20.9, C/P ratio of 17.65, OM of 20.1%, humidity of 71.4%; Ca of 2.43%, Mg of 0.26%, S of 0.39%, P (H2O) of 0.33%, P (CNA-+-H2O) of 0.40%, P (citric acid) of 0.40%; P2O5 (total) of 0.98% and K of 0.25%. In Thailand, Meunchang et al. found pH (water 1:5) of 7.7, C/N ratio of 14, OM of 48%, P (total) of 0.96%, K of 0.39%, Ca of 7.1%, Mg of 0.4%, Cu of 1.9-mg/kg, Zn of 51-mg/kg, Mn of 257-mg/kg, and Fe of 803-mg/kg.

The filter cake can be applied in the entire area at pre-planting in the furrow or planting lines. According to Nunes Júnior, 20-t/ha of filter cake on wet base or 5-t/ha on dry base can provide up to 100%

of the N required by the plants and 50% of the P, 15% of K, 100% of Ca, and 50% of Mg. In ratoon cane, the application of 70-t/ha of filter cake provided the greatest productivity of sugarcane stalks, regardless of application mode.

Coffee

Coffee is the second most traded commodity in the world, behind oil. The coffee husk is a common residue of the processing on farms, accounting for about 50% of the dry fruit harvested. This residue can return to coffee plants for nutrients release, providing around 29-g/kg of N (mostly in the form of nitrate) and 45-g/kg of K .Values of 23-g/kg of K-total (7.4-g/kg K-soluble), 14.8-g/kg of N, C/N ratio of 30, and 21-g/kg of lignin have also been found in coffee husks naturally dry. The K release from husks is high (above 90%) and regardless of the coffee bean constitution, decomposition rate, or processing type.

Another processing variation is the pulping done by the wet method, which withdraws just the husks, and the pulp and grain are placed to dry along with the parchment. Afterward, the parchment is removed from the coffee bean, which constitutes 12% of the dry fruit harvested. The pulp and parchment have 3.65 and 0.38% of K and 1.85 and 0.59% of N, respectively. The C/N ratio in the pulp is around 24 while in the parchment is 63. In the husk in the same processing, 38.9-g/kg K-total (17.7-g/kg K-soluble), 26.7-g/kg of N, C/N ratio of 16 and 20.9-g/kg of lignin were found. The variation in nutrient content within the same post-harvest processing must be taking into account, but mainly when they are from different processing (Figure 3). Composting is also a good option for husk use.

Figure: *Coffee pods stack after processing. This residue returns to farming by providing mainly K and N. Photo by Lucas de Ávila-Silva.*

Eucalyptus

Eucalyptus crops with ever shorter cycles (6-7-years) are extremely dependent on the biogeochemical cycling of nutrients from the residues that are added to the soil over the cycle and also at harvest. A eucalypt forest produces an average of 75% of commercial wood, 1.5-3% of leaves, 4-6% of branches, 6-19% of barks, and 10-12% of roots, considering the total biomass, after 7-years.

Studies have shown that eucalyptus planted forests can deposit 7-84-tons of dry matter to the soil during 7-years, from old dead branches and dry fruits (25-30%), barks (10-15%) and leaves (55-65%).

The treetops begin to close between the 1st and 2nd year after planting, and the competition causes the disposal of branches and lower leaves that are gradually deposited on the ground. The trees are taller and with small treetops from the 3rd to the 4th year, occurring the deposition of barks. However, the quantities of residue deposited to the ground depend on the eucalyptus species, climate, and evapotranspiration. The components of the eucalyptus forest deposited to the ground are called litterfall, which has great influence on the nutrients availability to the eucalypt.

Thus, considering that the estimated content of nutrients accumulated (in relation to the total accumulated in the plant) in the tree tops and eucalyptus bark after 6.5-years has on average 65% of N, 70% of P, 64% of K, 79% of Ca, and 79% of Mg, considerable amounts of nutrients are deposited to the ground and are considered in the fertilization management. In addition, the harvest of trees leaves in the area large amounts of residues such as leaves, branches, tree tops, and small trees discarded during harvest. The trees can be pruning and strips in the area or at companies, depending on the harvesting modules used. Some companies separate the so-called woody debris (thick branches, tree tops and small trees) and sell as wood or transformed into wood chips to produce biomass fuel for the company itself, depending on the demand.

However, we must emphasize here the importance of the retention of crop residues in the area because during the harvest the accumulated litterfall on the soil surface has order values of 8-14-t/ha. Studies have been shown that lower amounts of nutrients are required in fertilization when the bark is left in the field at harvest. Moreover, the roots also remain in the area, since that the currently practice of stump removal is increasingly scarce due to the large impact on the soil.

The crop residues have nutrient availability potential remaining in the area and can reduce the impact on the soil due to the heavy-machinery used. This is an important fact, since forestry operations can alter the physical and mechanical properties of the soil, increasing soil compaction. The productivity of eucalyptus forests may reduce with increasing soil compaction levels, due to: (i) physical obstruction of developing roots; (ii) lower water and nutrients absorption; (iii) gas exchange reduction.

Many factors influence the residue decomposition rates in the soil, with later nutrients availability for plants as we mentioned in the item 2. Generally, the leaves have faster biodegradation, since they have C/N ratios of 25-45 and C/P of 250-300, while the branches and trunk have C/N of 350-500 and C/P of 500-700 and the barks have C/N of 150-250 and C/P of 300-450. However, the nutrients allocated in the residues will not be fully available for eucalyptus plants; thus, many companies conduct tests in their crop areas trying to estimate the decomposition rates of the plant compartments and the recovery rates of applied fertilizers. Many companies use these data to establish the complementary fertilization management, seeking to minimize costs and make a more sustainable system.

Works describing the decomposition rates of different compartments in different environments can be found in the literature, which can be used for fertilization estimates when perform the tests is not possible. Some modeling programs such as the Nutricalc-UFV, allows us to estimate the different rates that will assist in the fertilization management. The residues have great importance on the nutrient balance of a system, and we cannot ignore the benefits of the construction and permanence of soil organic matter.

Chapter 7

Biofertilizers: Role and Management

Introduction

A biofertilizer is a substance which contains living micro-organisms which, when applied to seeds, plant surfaces, or soil, colonize the rhizosphere or the interior of the plant and promotes growth by increasing the supply or availability of primary nutrients to the host plant. Biofertilizers add nutrients through the natural processes of nitrogen fixation, solubilizing phosphorus, and stimulating plant growth through the synthesis of growth-promoting substances. The micro-organisms in biofertilizers restore the soil's natural nutrient cycle and build soil organic matter.

Figure: *Blue-green algae cultured in specific media. Blue-green algae can be helpful in agriculture as they have the capability to fix atmospheric nitrogen to soil. This nitrogen is helpful to the crops. Blue-green algae is used as a biofertilizer.*

Through the use of biofertilizers, healthy plants can be grown, while enhancing the sustainability and the health of the soil.

Biofertilizers can be expected to reduce the use of synthetic fertilizers and pesticides, but they are not yet able to replace their use. Since they play several roles, a preferred scientific term for such beneficial bacteria is "plant-growth promoting rhizobacteria" (PGPR).

Biofertilizers Today

Biofertilizers provide "eco-friendly" organic agro-input. Biofertilizers such as Rhizobium, Azotobacter, Azospirilium and blue green algae(BGA) have been in use a long time. Rhizobium inoculant is used for leguminous crops. Azotobacter can be used with crops like wheat, maize, mustard, cotton, potato and other vegetable crops. Azospirillum inoculations are recommended mainly for sorghum, millets, maize, sugarcane and wheat.

Blue green algae belonging to the general cyanobacteria genera, Nostoc, Anabaena, Tolypothrix and Aulosira, fix atmospheric nitrogen and are used as inoculations for paddy crop grown both under upland and low-land conditions. Anabaena in association with water fern Azolla contributes nitrogen up to 60 kg/ha/season and also enriches soils with organic matter.

Seaweeds are rich in various types of mineral elements (potassium, phosphorus, trace elements etc.) hence they are extensively used as manure by people of coastal districts. Seaweed - manure also helps in breaking down clays. Fucus is used by Irish people as manure on a large scale. In tropical countries bottom mud of dried up ponds which contain abundant blue green algae is regularly used as manure in fields. The mixture of seaweeds and blue green algae may serve as ideal fertilizer.

Phosphate-solubilizing Bacteria

Other types of bacteria, so-called phosphate-solubilizing bacteria, such as Pantoea agglomerans strain P5 or Pseudomonas putida strain P13, are able to solubilize the insoluble phosphate from organic and inorganic phosphate sources.In fact, due to immobilization of phosphate by mineral ions such as Fe, Al and Ca or organic acids, the rate of available phosphate (Pi) in soil is well below plant needs. In addition, chemical Pi fertilizers are also immobilized in the soil, immediately, so that less than 20 percent of added fertilizer is absorbed by plants. Therefore, reduction in Pi resources, on one hand, and environmental pollutions resulting from both production and applications of chemical Pi fertilizer, on the other hand, have already demanded the use of phosphate-solubilizing bacteria or phosphate biofertilizers.

Benefits

1. Biofertilizers are means of fixing the nutrient availability in the soil. Generally Nitrogen deficiencies.
2. Since a bio-fertilizer is technically living, it can symbiotically associate with plant roots. Involved microorganisms could readily and safely convert complex organic material into simple compounds, so that they are easily taken up by the plants. Microorganism function is in long duration, causing improvement of the soil fertility. It maintains the natural habitat of the soil. It increases crop yield by 20-30%, replaces chemical nitrogen and phosphorus by 30%, and stimulates plant growth. It can also provide protection against drought and some soil-borne diseases.
3. It has also been shown that to produce a larger quantity of crops, biofertilizers with the ability of nitrogen fixation and phosphorus solubilizing would lead to the greatest possible effect
4. They advance shoot and root growth of many crops versus control groups. This can be important when implementing new seed growth.
5. Biofertilizers also promote healthy soil, leading to greater farming sustainability.

Groups of biofertilizers

1. Azolla-Anabena symbiosis: Azolla is a small, eukaryotic, aquatic fern having global distribution. Prokaryotic blue green algae Anabena azolla resides in its leaves as a symbiont. Azolla is an alternative nitrogen source. This association has gained wide interest because of its potential use as an alternative to chemical fertilizers.
2. Rhizobium: Symbiotic nitrogen fixation by Rhizobium with legumes contribute substantially to total nitrogen fixation. Rhizobium inoculation is a well-known agronomic practice to ensure adequate nitrogen.
3. Streptomyces grisoflavus
4. Unigrow (UniGrow): a commercial bio fertilizer that is currently in use. It is made with a by-product of palm oil production and it contains a microbial element It has been shown to have promising results in studies.

Areas in Need of Improvement

Biofertilizers have been shown to have varying effects in different environments, and even within the same environment. This is something that many scientists have been working on, however there is no perfect solution at this time. They however, have been shown to have the most profound effects in drier climates.[6] In the future, it is hoped that biofertilizers effects will be better controlled and regulated in all environments.

Mechanism of Bio-fertilizers for Improving Soil Health

Role of Biofertilizers in Increasing the Soil Organic Carbon

The application of bio-fertilizers that contain living microorganisms is one of the management practices that can help to maintain or increase the content of organic matter and improve soil fertility in arable soils. the influence of bio-fertilizers on the acceleration of the humification of fresh organic matter that is introduced into the soil, that is why, especially when a large amount of exogenous organic matter (e.g., natural fertilizers or different bio-wastes) is introduced into the soil, the application of specially composed biofertilizers is very important in order to accelerate the transformation of the biomass. Nutrient Transformation Biofertilizers comprise a promising tool in agricultural ecosystems as a supplementary, renewable and eco-friendly source of plant nutrients. As they have an ability to transform nutritionally important elements from nonusable to highly assimilable forms without deleterious effects on natural environment, they are an important component of Integrated Plant Nutrient System. Application of biological fertilizers is thought to be a key element in maintaining soil fertility and crop productivity on the sufficiently high level, indispensable to achieve sustainability of farming. Biological Nitrogen Fixation Biological N2 fixation (BNF) is an important biochemical process associated with plants kingdom. The BNF is carried out by certain bacteria, actinomycetes, blue green algae, fungi and yeast. The most important of these are the bacteria, which have the ability to fix the atmospheric N- in symbiosis with the plants or as free-living organisms (non-symbiotically). Symbiotic N fixation is of more importance and the amount of N fixed is quite appreciable.

Among these, symbiotic (Rhizobia species) and nonsymbiotic fixers of N (Azotobacter and Az spirillum) are commonly used in agriculture. Free living blue-green algae in rice fields can provide 10-20 kg N/ha, while azolla, a fern, in association with blue-green algae, can supply

30-40 kg N/ha. However, these biofertilizers alone cannot meet the N requirement of crops and their use has to be integrated with fertilizer N to get a good yield. Role in Solubilizing Phosphorous Application of PSM by inoculating in soil appears to be an efficient way to convert the insoluble P compounds to plant-available P form, resulting in better plant growth, crop yield, and quality. Bacillus, Pseudomonas, Rhizobium, Aspergillus, Penicillium, and AMR are the most efficient P solubilizers for increasing bioavailability of P in soil. PSM provokes immediate plant growth by providing easily absorbable P form and production of plant growth hormones such as IAA and GA. Potassium Mobilizer Some of the soil microorganisms are well known for the solubilization of potassium from K minerals. Most of the micas, i.e., orthoclase, mica, illite, and muscovite, are the source of K in soil; with the reaction of microorganisms, produced organic substances solubilize the K and enhanced the soil solution, these organisms also produce various types of amino acids, growth-promoting compounds (IAA and gibberellic acid), and vitamins, which promote the crop growth and yield. Frateuria aurantia is well known for the mobilization of K from the mica; it also works in association with other biofertilizers in positive mode and enhances the soil productivity and fertility. Zinc Mobilizer Across the globe, Zn is known as one of the micronutrients and its deficiency limits the growth of crop and adversely affects the crop yield. The Zn fertilizers are more costly and its availability is also limited. In this regard the use of Zn solubilizers can play a vital role in enhancing the yield and quality of crops. Some of the microorganism species are isolated, which are responsible for Zn solubilization, i.e., Bacillus sp. The external applications of Zn fertilizers, soluble zinc sulphate (ZnSO4), are more common in all crops. The applied soluble Zn fertilizers during crop get into plant available form after reaction with soil constituents. Acid soil cation exchange and high-pH soil chemisorption's (Zn-CaCO3) or complexation by organic ligands fix the Zn in soil and reduce the concentration in soil solution. The microorganisms that are well known for the solubilization of Zn are B. subtilis, Thiobacillus thiooxidans, and Saccharomyces sp. These strains are used for Zn biofertilizers and get positive response among the farmer communities.

Effect of Biofertilizers In Soil Properties

It can be said that in general the application of bio-fertilizer gave significant effect to almost all of soil parameters. This condition can happen because the symbiotic relationship between AM fungi and a variety of plants that can produce colonies on the outside part in a root system. This condition can make the uptake of water and nutrients by the plant roots increases. The AM fungi can improve plant performance under drought stress through the increase in absorption of water and some nutrients.

That are zinc (Zn) and copper (Cu), and also plant variables like a leaf height, leaf water turgidity, stomatal activities, and root growth. The AM fungi performance as an agent which can improve plant-water relationship through increasing stomatal resistance by adjusting plant hormonal balance. Moreover, through this chain, the P element can increase by the activity of AM fungi in a growth phase. The AMF symbiosis can also increase the absorption of other nutrients such as P, N, Cu and Zn. Beside that process, the additional material combined with biofertilizer (growing substrate, humic acid and Phyto-compound) can make the performance from this product much better.

This is supported, that humic acid has a role in the release of P adsorbed in the soil and can increase the availability of P in the soil. Humic acid can replace phosphate ions by sorption's mechanism and also has the ability to binding the organic compounds. This process can happen because the negative charge on the functional groups in the humic acid has the ability to react and interact with positively charged ions. These drawbacks are combined with a high cost of production that is beyond the means of many farmers in the developing world. The application of biofertilizers is eco-friendly, relatively inexpensive, nontoxic, and possesses the significant potential to increase plant yield.

Thus, the function of plant growth-promoting microorganisms and the application of biofertilizer made from viable microbial strains to the field bodes well for successful management of the rhizosphere for sustainable agriculture.

Chapter 8

Integrated Nutrient Management in Cropping Systems

Introduction

Recently, most agrarian countries have witnessed either declining or stagnant crop yields. Inadequate soil organic matter (SOM) due to the poor physical, chemical, and biological properties of the soil leads to an overall decline in the productivity of farmlands. Therefore, the adoption of integrated nutrient management (INM) practices is vital to revive sustainable soil health without compromising yield potential. Integrated nutrient management is a modified nutrient management technique with multifarious benefits, wherein a combination of all possible sources of plant nutrients is used in a crop nutrition package.

Several studies conducted in various parts of the world have demonstrated the benefits of INM in terms of steep gain in soil health and crop yields and at the same time, reducing greenhouse gas emissions and other related problems. The INM practice in the cropped fields showed a 1,355% reduction in methane over conventional nutrient management. The increase in crop yields due to the adoption of INM over conventional nutrient management was as high as 1.3% to 66.5% across the major cropping systems. Owing to the integration of organic manure and residue retention in INM, there is a possibility of significant improvement in soil aggregates and microbiota. Furthermore, most studies conducted to determine the impact of INM on soil health indicated a significant increase in overall soil health, with lower bulk density, higher porosity, and water-holding capacity. Overall, practicing INM would enhance soil health and crop productivity, in addition to decreasing environmental pollution, greenhouse gas emissions, and production costs. The increasing human population and their consequent need for food, combined with the depletion of healthy soil, have led to unprecedented damage to natural resources, making it difficult to meet the global demand for food. In the developing world, achieving food security through sustainable systems is a big task, yet it is vital for poverty alleviation.

To get around this problem, farmers have resorted to overusing specific inputs like chemical fertilizers and pesticides, which have already begun to harm the ecosystem. During the initial days of fertilizer usage, the impact of crop fertilization with inorganic fertilizers has been prominent in world agriculture (Hossain and Singh, 2000). In most countries of agrarian background, manufacturing and service of chemical fertilizers have been practiced as prime agenda in securing nations' food and nutritional security. India, a populous agrarian country, is the world's third-largest producer and consumer of chemical fertilizers (Tandon and Tiwari, 2007). As per recent reports, the Indian fertilizer market reached a value of Rs. 887 billion and is expected to grow at a compound annual growth rate of 5.5% by 2026 [IFA (International Fertilizer Association), 2020]. Food production must increase significantly while agriculture's environmental impact must decrease greatly to fulfill the world's future food security and sustainability needs (Foley et al., 2011).

India is a country with diverse cropping practices. The nutrient mining by various crops and cropping systems is far higher than the nutrient additions annually through fertilizers (Kumar et al., 2015). For the past 40-years, a nutrient gap of 8 to 10 million tons of nitrogen (N), phosphorus (P), and potash (K) per year have been documented (Tandon, 2004). This condition is analogous to depleting the soil's nutritional reserve. Long-term negative impacts of imbalance and indiscriminate use of inorganic fertilizers, particularly NPK-based formulations, have irreversibly damaged the soil resource base of many agroecosystems. Despite the tremendous improvement in crop productivity in numerous crops due to improved varieties and increased use of agrochemicals, the goal of ensuring food and nutritional security remains challenging (Nath et al., 2018). The ever-increasing food demands of the burgeoning population have continuously exerted pressure on the agroecosystems. In the process of increasing agricultural production, the agriculture production units significantly contribute to environmental pollution. For future generations, the goal is no longer to enhance agricultural production but also to optimize nutrients, energy, and water usage while minimizing environmental impact. Over-exploitation of nutrients from the soil and poor nutrient loss replenishment, depleted nutrients from the soil are often unable to be replenished by artificial crop fertilization, resulting in an imbalance in the soil nutrients pool. Hence, the huge increase in global greenhouse gas (GHG) emissions by the agriculture sector, primarily due to the use of synthetic fertilizers and pesticides in recent years, both of which are rapidly expanding. One of the primary causes of environmental pollution, such as eutrophication and GHG emissions is the injudicious use of fertilizer, particularly nitrogen (Davidson et al., 2014). Thus, it is high time to search for innovative practices that guarantee higher yields with the least amount of additional environmental damage possible, especially for developing countries.

No single source of plant nutrients, such as chemical fertilizers, organic manures, crop residues, and bio-fertilizers, can meet the entire nutrient need of a crop in today's intensive agriculture systems (Mahajan and Gupta, 2009). In this context, the integration of all possible sources of nutrients is ideal for enhancing the soil resource base besides contributing to crop productivity (Nath et al., 2018). Findings of various studies suggest integrated nutrient management (INM) is a tool that can offer good options and economic choices to supply macro and micronutrients of plants and also contribute to reducing the dependence on externally purchased chemical fertilizers besides protecting soil health. Physical properties related to soil structure, are greatly influenced by adding organic manures (Das et al., 2014).

An increasing number of research have suggested that INM has an impact on crop production, soil quality, and the environment while balancing food security and GHG mitigation. Numerous earlier studies largely examined how INM affected crop productivity, soil quality, or environmental performance. It is currently unknown how INM will affect crop productivity, soil bulk density, microbial biomass carbon, and the environment as a whole. This review addresses the synergistic effects of different nutrient sources and their combinations. This includes exploring the interactions between organic and inorganic fertilizers, biofertilizers, and crop residues on crop productivity and soil quality. By understanding how different nutrient sources interact, the researcher can develop more effective nutrient management strategies that optimize nutrient use efficiency and minimize environmental impacts. Further, this review highlights the long-term impacts of INM on soil quality parameters such as SMBC, bulk density, and dehydrogenase activity. This review article helps in understanding how INM affects soil quality, productivity, and environment. The researchers can develop more sustainable soil management practices that protect and improve soil health, and can develop more effective fertilizer recommendations and minimize nutrient losses. Finally, we can help farmers and policymakers make informed decisions about nutrient management.

Fertilizer Consumption in India and the Need for INM

Global use of inorganic fertilizers have increased almost fivefold since 1960 and have significantly supported population growth (FAO, 2017). The present world's demography will change in many folds in the near future and will require the production of 70% extra food, fodder and fuel to meet the demands of the ever-growing population (FAO, 2017). On the contrary, the natural base resource required for additional food production is shrinking. A study conducted by FAO (2017) indicated that the growth of agriculture between 1960 and 2015 exhibited about a 28% increase in the production of 174 crops cultivated all over the world. This phenomenal gain in production over the past few decades has not only tripled production but also paved the way for land degradation by hampering the soil's physical, chemical, and biological resource base (Power, 2013).

High-input-driven and resource-intensive farming practices have caused massive and irreparable damage to soil and water resource in the agro-ecosystem and deteriorations in soil health besides contributing enormously to GHG emissions. Among agricultural practices, nutrient management is one such practice that plays a critical role in crop productivity and soil health. Additionally, after the green revolution, an over-reliance on fertilizers led to a decline in the efficiency of nutrient use and an increase in GHG emissions. The use of fertilizers can have implications for the accumulation of heavy metals in the soil and plants, which can subsequently enter the food chain, leading to potential pollution of water, soil, and air. Some mineral/chemical fertilizers contain low quantities of heavy metals and radionuclides, and their excessive application in agriculture can pose environmental problems.

For example, the excessive use of urea has been identified as a major contributor to increased nitrate (NO3-) levels in drinking water and river systems. Studies in agricultural districts of Srikakulam in Andhra Pradesh, located in the Vamsadhara river basin, revealed elevated nitrate concentrations ranging from traces to 450-mg NO3-/L of water, particularly following fertilizer applications (Rao, 2006).

The current consequences of nitrate pollution in freshwater bodies in India reflect the detrimental effects of historical and ongoing excessive use of fertilizers and manures. The transport of phosphatic fertilizers through surface water flow can also contribute to increased phosphate content in drinking water and rivers. The rising levels of dissolved N and P loads are primarily associated with increased losses from agricultural and sewage systems.

Moreover, the excessive application of nitrogen fertilizers has implications for GHG emissions. The relative contribution of fertilizer nitrogen application to total GHG emissions has increased from 8% in 1970 to 23% in 2010. NH3 volatilization from rice fields is higher when N fertilizers are surface-broadcasted, leading to reduced nitrogen use efficiency (Ladha et al., 2005). Additionally, excessive N fertilizer application can negatively affect soil health by degrading soil carbon and affecting the structure and function of soil biological communities. Long-term application of N fertilizer alone has been shown to significantly reduce soil pH at various experimental sites in India.

Further, phenomenal loss in nutrient use efficiency of various crops and cropping systems has gradually encouraged farmers to apply higher doses of nutrients. As a result, this injudicious practice of nutrient management has paved the way to impair soil health due to less or no use of organic manure, residue retention, and bioinoculant-mediated fertilizer management.

This has strongly distracted the complementarity between bio-geo cycles of the agroecosystem and is the main reason for multiple nutrient deficiencies, declining fertilizer response, and crop productivity.

To address the current challenges associated with excessive fertilizer use, including the potential accumulation of heavy metals, pollution of water sources, GHG emissions, and adverse effects on soil health. Implementing sustainable nutrient management practices that optimize fertilizer use, minimize losses, and protect both the environment and human health is imperative. Hence, INM is necessary to bring back harmony in the agroecosystem besides sustaining the productivity of crop and soil health.

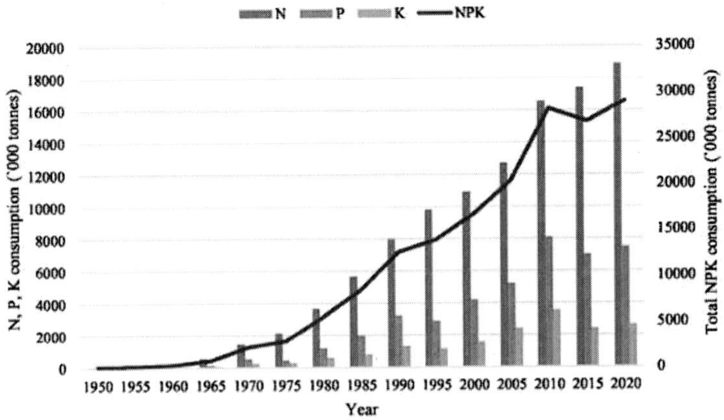

Figure: *Annual fertilizer consumption in India (Fertilizer Association of India, 2021).*

The INM primarily refers to the judicious, efficient, and integrated use of all available sources of organic, inorganic, and biological components to combine traditional and modern techniques of nutrient management into an environmentally sound and economically optimal agricultural system (Janssen, 1993). To synchronize nutrient demand by the crop and its release in the environment, it optimizes all elements of the nutrient cycle, including N, P, K, and other macro- and micronutrient inputs and outputs. The INM techniques reduce losses due to leaching, runoff, volatilization, emissions, and immobilization while maximizing nutrient use efficiency (Zhang et al., 2012). Additionally, INM aims to improve the physical, chemical, biological, and hydrological aspects of the soil to increase agricultural production and reduce land degradation (Janssen, 1993; Esilaba et al., 2005).

There is now a deeper understanding that INM may concurrently and very invisibly protect soil resources while also increasing crop yield. Farmyard manures, farm wastes, soil amendments, crop residues, chemical fertilizers, green manures, cover crops, intercropping, crop rotations, fallows, conservation tillage, irrigation, and drainage are all used in its methods to increase plant nutrition and preserve water (Janssen, 1993).

The INM practice also promotes methods designed to reduce nutrient losses and enhance plant uptakes, such as the deep placement of fertilizers and the use of inhibitors or urea coatings (Zhang et al., 2012). Instead of merely concentrating on yield-scaled profit, these techniques urge farmers to focus on long-term planning and offer scope for the reduction of environmental impact.

Nutrient mobilization/mineralization in soil under INM practice

Farmers generally apply significant amounts of N fertilizer at the time of sowing or planting under conventional practice. Typically, 80% of the total N fertilizer is applied as a basal dressing, and the remaining N is applied within the first 10-days after transplanting for rice or within the first 30-days after the date of seeding for wheat, maize, and sorghum. Due to the inefficient synchronization between the soil's supply and crop demand caused by this N application pattern, a significant amount of inorganic N is available in the soil before it is needed by rapidly growing crops (Chen et al., 2006). However, a material with a high N content (fertilizers) favours net mineralization, whereas one with a low N concentration (manures) results in net immobilization. Interestingly, Reddy et al. (2008) stated that organic manures (farmyard manure-FYM) with a C:N ratio of 29% to 33% immobilized N during the initial 40-days of incubation. This caused net N mineralization to begin during the first week of incubation, and then urea fertilizer has been added to lower the C:N ratio to 18%-22%.

Therefore, a rapid increase in the mineral N near the end of incubation was the result of the combined application of fertilizer with FYM. The combined usage of organic manures with that of chemical fertilizer assures the continuous supply of nutrients by acting as a multi-nutrient pool, further during organic matter mineralization nutrients release slowly and gradually so that the crop enjoys the soil nutrient pool throughout its requirement. Consequently, the continual application of inorganic fertilizers in combination with FYM or lime, significantly altered soil microbial biomass carbon, soil N and P, fulvic acid (FA), and humic acid (HA; Srinivasarao et al., 2020). In addition, integrated use of fertilizer, manure, and lime application in soybean-wheat rotation led to improvements in soil water retention, soil aggregates, microporosity, and water holding capacity as well as a decrease in the soil's bulk density (BD) in the top 30-cm of the soil when compared to fertilizer application alone (Hati et al., 2008). Thus, the use of organic manures enhances fertilizer use efficiency and serves as an alternative source of nutrients (Dwivedi et al., 2016). The INM system synchronizes the nutrient demand set by plants, both in time and space, with the supply of nutrients from the labile soil pool and applied nutrient sources (Cassman et al., 2002). As INM encourages the split application of N fertilizers during critical stages of crop growth in small quantities it has the potential to increase crop productivity and quality and also reduce nutrient losses (Tilman et al., 2002; Witt and Dobermann, 2004).

Data collection and rice equivalent yield The key data collected from the publications that qualified for review included grain yield, soil bulk density (BD), SOC, soil microbial biomass carbon (SMBC), dehydrogenase (DHA), GHG emission. In this analysis, change (-) in measured variables between the conventional and INM practice was expressed as a percentage, i.e., --= -100*(INM - Conventional)/Conventional because of its ease of interpretation. To compare different monocrops and cropping systems, rice equivalent yield (REY; equation 1) was determined by converting the economic yield of different crops on the basis of their marketable price prevailing during the period for each crop, including rice, and expressed in ton per unit area.

Effect of INM on Crop Productivity, Soil bulk Density, Microbial Biomass carbon, and Reducing Environmental impact

Crop yield: Yield is the utmost concern in agriculture production systems. To achieve a desirable yield, a plant nutrient management system is one of the prime agro-practices. The impact of integrated nutrient management on rice equivalent yield was synthesized by taking into consideration of yield obtained under INM-treated plots against the conventional nutrient management of various crops and cropping systems and presented in Table 1. The results of rice equivalent yield indicated the gain in productivity of most of the cropping systems. The increment in crop yields under INM over conventional nutrient management across the cropping system studied was 1.3% to 66.5%. Furthermore, significant improvement in crop yields was more noticeable in field crops, especially rice, wheat, and soybean. However, vegetable crops like okra, tomato, and onion crops also demonstrated increased yields under INM. A long-term field experiment in India from 1973 to 2004, observed a substantial decrease trend in soybean yield for both the control and NPK treatments whereas the NPK-+-FYM treatment had a statistically significant (p-<-0.05) increase in yield over time. Increased microbial activity, better supply of macro- and micronutrients like S, Zn, C, and B, which are not supplied by NPK (straight) fertilizers, and lower nutrient losses from the soil are some of the additional benefits of organic matter over N, P, and K supply that may have contributed to the higher yields of soybean and wheat obtained with the FYM-+-NPK treatment. Therefore, INM system enhances the yield potential of crops over and above achievable with recommended fertilizers. In the future, there are several promising areas of work that can be explored in INM to further enhance crop yield. One avenue is the development and utilization of precision nutrient management techniques. As precision nutrient management uses advanced technologies such as remote sensing, geospatial analysis, and machine learning algorithms to assess crop nutrient requirements at a fine-scale level and deliver precise and site-specific nutrient applications. This approach minimizes nutrient loss and ensures that crops receive nutrients when and where they are most needed. Another important issue to be addressed is the exploration of microbial interventions in nutrient management. Microbial interventions in nutrient management use beneficial microorganisms like mycorrhizal fungi, rhizobacteria, and other plant growth-promoting microbes to improve nutrient availability and uptake efficiency, and enhance crop productivity. Scope of INM should be extended to the use cover crops and crop rotations to build soil fertility, enhance nutrient cycling, and crop yield.

Greenhouse gas emission: Globally, agriculture and its associated systems are viewed as a massive contributor of GHG emissions (Ravikumar et al., 2021). To reduce its contribution, reduction in excess nutrient application and balanced application are the key mitigation strategies (Sapkota et al., 2021). Enhancing crop yields by application of plant nutrients become an ambiguous practice in most farmlands (Garnett et al., 2013). On the contrary, the imbalanced application of fertilizer in croplands is a major source of anthropogenic greenhouse gas emissions. Therefore, to keep these anthropogenic GHG emissions under control, devising, and practicing proper fertilizer management is essential (Carlson et al., 2017). Of the several means of nutrient management, INM is proven better at minimizing GHG emissions. The work conducted by Nyamadzawo et al. (2017) in Zimbabwe in maize indicated practicing INM reduced about 17.1% GHGs over conventional nutrient management. Another study in China demonstrated a significant (p-<-0.05) reduction (1,268%) in GHG emissions against conventional nutrient management in the rice-wheat cropping system. Similarly, several studies conducted in India indicated a considerable reduction in GHG emissions (20%-1,355%) in rice, wheat, and mustard crops.

Uncontrolled use of fertilizers increases emissions into the atmosphere and groundwater leaching of nutrients. The INM promotes high agricultural yields while reducing N losses and associated detrimental consequences on the environment (Gruhn et al., 2000). The ultimate fate of applied fertilizers is a combined effect of crop nutrient intake, immobilization, and soil residues, as well as nitrogen losses to the environment as ammonia volatilization, NOX emissions, denitrification, N leaching, and runoff (Wu and Ma, 2015). Additionally, the pattern of N application, crop features, soil characteristics, climate, and management approaches affect the efficacy of applied fertilizers. INM thus recommends deep urea placement, which can greatly boost N-use efficiency with low NH3 volatilization and reduces nitrate-leaching (Jambert et al., 1997). Because nitrification occurs mostly following fertilizer N application (Ma et al., 2010) and irrigation, the use of nitrification inhibitors can also minimize N2O emissions (Ju et al., 2011). Additionally, INM supports the use of organic nutrient sources since they provide both greater potential for agriculture's sustainability and more immediate environmental advantages. Combining organic manure with other management techniques, such as incorporating agricultural residues and creating conservation tillage (such as no-till or reduced-tillage practices), can also help to lower GHG emissions, enhance soil quality, and boost carbon sequestration (Huang and Sun, 2006).

Reducing environmental pollution can be achieved by developing precision nutrient management techniques, promoting nutrient recycling and reuse strategies, exploring the use of cover crops and diversified cropping systems. Precision nutrient management techniques minimize nutrient losses to the environment by refining nutrient application methods, such as incorporating controlled-release fertilizers or using site-specific technologies to target nutrient placement. In INM, nutrient recycling and reuse strategies efficiently capture and recycle nutrients from various sources, such as agricultural residues, livestock manure, and wastewater. Practices like anaerobic digestion, composting, and biochar production can be used to convert these nutrient-rich materials into valuable organic amendments. Intercropping and crop rotation can also be made an integral part of INM systems that incorporate nitrogen-fixing crops can also reduce the need for synthetic nitrogen fertilizers and decrease nitrogen runoff.

Carbon sequestration: Soil OM is the most important indicator of soil fertility, quality, and productivity, and is usually estimated by determining SOC. It is generally believed that fertilizer application increases residues, including roots, returned to the soil, and as a result, can increase SOM content and C sequestration. The INM through green manuring, crop residue incorporation and other animal-based manures has a profound influence on soil carbon stock. From their study, Sujata et al. (2007) revealed application of inorganic fertilizers in combination with organic manure has a positive influence on all soil properties, especially soil organic carbon. Similar, observations were also made by Singh et al. (2009) where the addition of inorganic fertilizers with various organic manures in rice-wheat system enhanced the soil particle aggregation which in turn resulted in higher storage of soil organic carbon content. For instance, several studies conducted in India to know the effect of integrated nutrient supply on soil organic carbon status indicated significant gain in soil organic carbon, especially under intensive cropping systems like rice-wheat, maize-wheat (Paramesh et al., 2014, 2020), maize-mustard, and rice-groundnut.

The direct application of organic manure to the soil, which encouraged the development and activity of microorganisms, as well as improved root growth, which led to increased biomass output, crop stubbles, and residues, are both associated with increasing the amount of organic carbon in the INM. The increased carbon content of the soil have been caused by the eventual decomposition of these components. Additionally, the inclusion of FYM facilitated the synthesis of humic acid, which subsequently raised the soil's organic carbon content. Further, they opined that increased residue return, and minimum or zero tillage practices as the contributing factors to increasing SOC under INM practice.

A review of the long-term fertilizer experiment in China by Lu et al. (2009) observed linear relationships between the amount of N application and straw incorporation to soil C sequestration. Contrary to this belief, Khan et al. (2007) observed a decline in SOC after 50-years of chemical fertilizer application despite crop residue incorporation due to excessive N removal by crops to the tune of 60%-190%. The possible reasons for the decrease in SOC under continuous chemical fertilizer application were (i) acid forming ammonium fertilizers application delay soil C decomposition, (ii) enhancement in the activities of heterotrophic soil microorganisms that use C derived from crop residues, and (iii) decrease in C:N ratio lead to bacterial-dominated microbial communities, this further results in faster decomposition of SOC (Moore et al., 2003). Increased decomposition of SOC will result in more dissolved organic C, which can be lost through leaching.

Enhancing carbon sequestration through integrated nutrient management can be achieved by exploring soil management practices, investigating the role of plant-soil-microbial interactions, exploring the integration of agroforestry practices, developing precision nutrient management techniques, and using modeling tools and decision support systems. Soil management practices that promote carbon sequestration include the use of organic amendments, cover crops, and crop rotations. These practices can help to increase soil organic matter, which is a major sink for atmospheric carbon. Beneficial soil microorganisms, such as mycorrhizal fungi, can help to improve nutrient uptake, promote root growth, and enhance soil aggregation. These activities can create conditions that are favourable for carbon sequestration. Agroforestry systems can provide multiple benefits, including increased biomass production and carbon storage. Precision nutrient management techniques can indirectly contribute to carbon sequestration by improving nutrient use efficiency and minimizing nutrient losses. Adaption of modeling tools can consider factors such as soil type, climate conditions, and crop management practices to estimate carbon sequestration rates and provide recommendations for optimizing nutrient management strategies that enhance carbon sequestration.

Soil Properties

Crops grow better under friable and well-aggregated soils with optimum soil bulk density as it greatly influencing crop root growth and nutrient uptake. Practicing long-term INM brings a favourable difference in soil bulk density (Saha et al., 2010). The main reason for decreasing bulk density under INM was aggregation of soil particles due to increasing organic matter as well as stability of aggregates which leads to an increase in the total pore space in the soil. Islam et al. (2012) also concluded that the addition of organic matter through organic manure decreases the bulk density of soil. Integration of organic sources of nutrients such as crop residue and organic manures has a significant effect on soil bulk density (Celik et al., 2004). Nevertheless, studies undertaken to know the influence of integrated nutrient management on soil bulk density indicated favourable and convincing results. A study conducted by Nayak et al. (2012) in rice-wheat cropping systems at various places indicated significantly lower values of soil bulk density in the plots treated with integrated nutrient management systems over conventional chemical fertilizers. Similarly, Saha et al. (2010) and Salahin et al. (2011) also revealed the positive benefits of integrated nutrient management on soil bulk density in maize-mustard and tomato-mung-toria cropping sequence (Table 4). A long-term fertilizer experiment conducted in a rice-wheat cropping system at Chattisgarh, India ascertained that the incorporation of organic sources with chemical fertilizer application decreased bulk density, increased infiltration rate, and available NPK status of the soil (Bajpai et al., 2006). Significant reduction of bulk density in INM may be due to better soil aggregation (Singh et al., 2000), higher organic carbon, and more pore space (Selvi et al., 2005). A similar reduction in bulk density of soil due to the application of FYM with 100% NPK was also observed by Bellakki et al. (1998) and Bhattacharyya et al. (2010).

Table 5 shows an increase in SMBC and DHA under INM over conventional chemical fertilization. The increase in microbial biomass is mostly driven by the microbial biomass found in the organic byproducts and the addition of carbon from the substrate, both of which encourage the naturally occurring soil microbiota. The combined effect of FYM and chemical fertilizers in raising the SMBC under the maize-wheat system was also highlighted by Verma and Mathur (2009). The application of biofertilizers is known to create a variety of growth-promoting compounds in addition to their basic effects, which help explain the rapid expansion of microbial growth. Additionally, the crops' rhizodeposition (Jones et al., 2009) contributed to the elevated SMBC under INM practice. Nayak et al. (2007) similarly demonstrated enhanced dehydrogenase activity with both compost and inorganic fertilizer application under continuous rice-growing situations and explained a significant relationship between SMBC. The addition of fly ash and FYM to a rice-wheat cropping system in Alfisol and Vertisol significantly enhanced the microbial biomass carbon and dehydrogenase activity in soil (Ramteke et al., 2017).

Kanchikerimath and Singh (2001) found that the inorganic fertilizer (NPK) increased crop yield, SOC, total N, mineralizable C and N, microbial biomass C and N, and dehydrogenase, urease, and alkaline phosphatase activities, while manure applied together with inorganic fertilizer increased these parameters more strongly. However, when more organic residues are added to the soil, they undergo microbial decomposition, which releases organic compounds like polysaccharides, which act as a strong binding agent in the formation of large and stable aggregates that help to improve the physical properties of soil (Manickam, 1993). Accordingly, Yuhui et al. (2004) highlighted the significance of organic manure addition with chemical fertilizers to improve yield, and soil health, and enhance the earthworm population in the soil. In another long-term experiment in Germany, Marhan and Scheu (2005) observed an increase in earthworm biomass by 42.8% in NPK-+-FYM treatment, with a decrease of 9.4% in NPK treatment. They ascertained an increase in earthworm biomass in NPK-+-FYM could be due increased utilizable SOM pool.

Future research in integrated nutrient management can prioritize enhancing soil quality, a critical aspect of sustainable agriculture. One promising area is the exploration of practices that enhance soil organic matter content and composition. Efforts can focus on identifying effective organic amendments, cover crops, and crop rotations that facilitate the accumulation of stable organic matter in soils. Understanding the interactions among nutrient management practices, organic matter dynamics, and soil microbial communities will provide valuable insights for optimizing soil quality enhancement.

Additionally, research should emphasize nutrient management strategies that promote soil nutrient cycling and enhance nutrient use efficiency. This involves investigating the synergistic relationships between organic and inorganic fertilizers, biofertilizers, and soil microorganisms in nutrient cycling processes. Developing nutrient management approaches that minimize nutrient losses through leaching, volatilization, and runoff, while simultaneously ensuring sufficient nutrient availability for crops, will improve nutrient use efficiency and contribute to soil quality enhancement. Moreover, such practices will help mitigate environmental pollution associated with nutrient runoff.

Integrated nutrient management on soil health and crop quality

The current review has shown a positive effect of Integrated nutrient management (INM) on yield, soil microbial activity, SOC, and BD over conventional chemical fertilizer application. These summarized results further highlighted reduction in use of chemical fertilizer under INM practices, this in turn resulting in reduction of environmental burden associated with fertilizer production, transport, and its use. The main objective of INM is to wisely utilize its three primary components.

These include harnessing the existing synergy between dual-purpose microbes (which promote growth and control soil-borne pathogens), limiting the use of chemical fertilizers, promoting the multiplication of native soil microbial diversity through organic substrates, and maintaining a nutrient inflow that exceeds outflow. Additionally, the aim is to ensure that production economics are favourable in the market. Nevertheless, there are still several crucial areas that require urgent attention to make INM a globally dynamic nutrient management approach. Through enhancement of SOC and soil microbial activity INM increases soil nutrient availability and it is having positive impact on crop quality.

Conclusion: Farmers have rich experiences in integrated and efficient utilization of different sources of organic materials to produce modest crop yields and maintain soil fertility using traditional farming practices. However, during the last two decades, to achieve food security with declining land and other resources, this practice is gradually being abandoned, and nutrient management is being shifted to over-reliance on chemical fertilizers. Over-application of N has become more common in intensive agricultural regions, leading to low nutrient-use efficiency and environmental pollution, which threaten the long-term sustainability of the agricultural system. Many factors might have contributed to the over-application problems including obtaining higher yields as a top priority, small land holding, lack of nitrogen management, and lack of effective extension systems. A review of experiments conducted globally indicates that chemical fertilizers alone are not enough to improve yield and soil quality at high levels. Further, the review also highlighted the negative effects of the continuous use of synthetic N fertilizers on soil organic carbon, bulk density, soil enzymatic activity, SMBC, and the environment. Crop yield responses to nutrient management may vary significantly from year to year due to variations in weather conditions and indigenous N supply, and thus the commonly adopted prescriptive approach to N management needs to be replaced by a responsive in-season management approach based on the diagnosis of crop growth and N status and demand.

Chapter 9

Soil and Fertilizer Management and Economics of Fertilizer Use

This chapter presents technologies for replacing the nutrients lost from cropped fields with external fertilizer sources in a manner that minimizes the consequences of too little or too much application. The technology of using industrial fertilizers, organic fertilizers, and application of farmyard manure and compost alone or in combination with industrial fertilizers are discussed in detail.

Introduction

Plants require nutrients for growth and reproduction. Most crops obtain these nutrients from the soil. Legumes obtain additional nitrogen from the air through biological nitrogen fixation. At harvest, nutrients contained in the crop products (grain, leaves, stalks, etc.) are removed from the fields. The quantity of nutrients removed varies with the crop species, yield, and the portion of the crop that is removed. Also, depending on management, topography, wind speed, soil moisture, and rainfall intensity, nutrients can be lost through soil erosion and leaching to deep soil layers, even below the level of the plant roots.

In East and Southern Africa (ESA), many soils are already limited in their capacity to supply nitrogen and phosphorus to plants, as well as being limited in potassium and secondary micronutrients (Kihara et al., 2017). Some nutrients may be present in limited quantities, while others are in sufficient supply; it is the nutrient(s) in low quantities that limit crop production (Figure 4.1). When plants grow, they take up nutrients from the soil, and the process of removing nutrients in crop products is known as nutrient mining. It is therefore important to replace the nutrients lost from the fields by adding more from external sources (such as fertilizers), while also protecting soils from the processes of erosion and leaching.

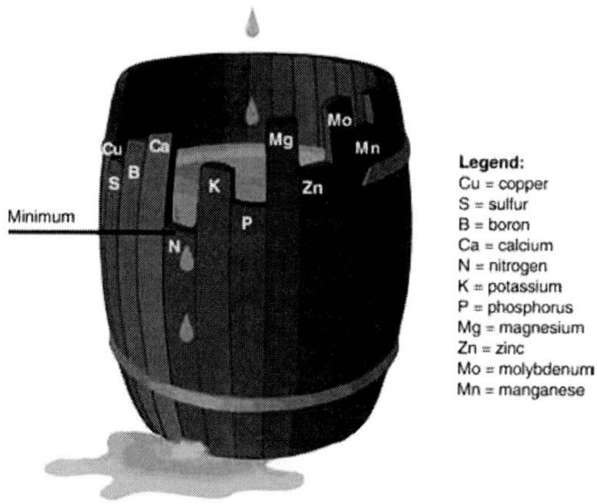

Figure: *Liebig's 'Law of the Minimum' adapted for soils of ESA, illustrating the importance of balanced nutrients for maximum plant growth and productivity. The yield of a plant is limited by a deficiency of any one essential element, even when all others are present in adequate amounts. In this illustration, too little nitrogen is the main problem. If nitrogen is added up to the top, phosphorus becomes the next limiting nutrient.*

When available nutrients are insufficient to meet crop requirements, yields will be below the maximum potential, and the long-term productivity of the land will decline due to nutrient mining. However, if nutrients are applied in excess of crop requirements, there is a risk of agronomic problems such as crop lodging (falling over) and poor grain yields due to excessive vegetative growth at the expense of grain production (in the case of excess nitrogen). Excess nutrients can also have negative effects on the environment, such as soil poisoning or toxicity, nitrate leaching into groundwater, phosphate runoff into surface water, and release of greenhouse gases to the atmosphere. Using too much fertilizer will also reduce the profitability of crop production.

Nutrients can be added to the soil by growing legumes in rotation or as intercrops, and by the application of industrial fertilizers, farm-made organic fertilizer (manure and compost), and other organic matter such as crop residues and biomass derived from cover crops and agroforestry trees. These technologies have a positive effect on the livelihoods of small-scale farmers by sustainably increasing yields, food security, and incomes.

Africa RISING research in ESA validated the use of industrial and organic fertilizers in a manner that would ensure good yields, and minimize the problems linked with application of too little or too much fertilizer. The trials took place in the highlands, medium altitudes, and lowlands of northern Tanzania (Babati district); the semi-arid zone of central Tanzania (Kongwa and Kiteto districts); and the semi-arid zones of central and southern Malawi (Machinga and Dedza districts).

Before making a decision about which fertilizers to apply, it is important to test the soil and identify which nutrients are lacking. The soils in Babati are of moderate fertility with deficiencies in nitrogen, phosphorus, and, in some locations, low levels of zinc. Soils in Kongwa and Kiteto are heavily eroded and of poor fertility, with widespread deficiencies in nitrogen, phosphorus, sulfur, zinc, and boron. Soils in Machinga and Dedza in Malawi are deficient mostly in nitrogen, sulfur, and zinc.

Industrial Fertilizers

Industrial fertilizers consist of inorganic or synthetically produced organic compounds containing nutrients for plant use. Organic compounds such as phosphate rocks that are further enriched with inorganic compounds are also considered as industrial fertilizers in the context of this chapter. These fertilizers are applied at planting, with nitrogen-containing fertilizers often used for additional top-dressing. The most common nutrients supplied in fertilizers are nitrogen, which promotes the growth of the plant, and phosphorus, which improves root growth and flowering. There are many types of industrial fertilizers; some contain one nutrient (straight fertilizers) while others contain two or more nutrients (compound fertilizers).

Africa RISING tested and validated the following types of fertilizers:

- Diammonium phosphate (DAP): Contains 18% nitrogen, 46% phosphate, and 1.6% sulfur. DAP is the world's most widely used phosphate fertilizer. It is popular because of its relatively high nutrient content and because it is water soluble. It temporarily increases soil acidity, but over time reduces it. Special care needs to be taken to avoid placing DAP near germinating seeds.
- Minjingu NAFAKA and Minjingu Mazao: Minjingu NAFAKA contains 9% nitrogen, 16% phosphate, 6% potassium, and a small quantity of calcium, magnesium, and sulfur, and the

micronutrients zinc and boron. This fertilizer is based on Minjingu phosphate, which is one of the most reactive rocks. Minjingu, mainly an organic fertilizer and often enriched with other chemical elements, is mined from Minjingu mines in the Manyara region of Tanzania, about 80 km from Babati. It is especially suited to soils that are deficient in phosphorus, and can be applied to acidic soils (pH >6) due to the liming effect of the calcium. Before enrichment with micronutrients, it was referred to as Minjingu Mazao.

- Nitrogen, phosphate, sulfur (NPS): Contains 23% nitrogen, 21% phosphate, and 4% sulfur; with a similar product containing zinc and potassium but with reduced phosphate (10%). This is suitable for areas where potassium is not a limiting nutrient. The sulfur included in NPS fertilizer helps the plants to develop chlorophyll for photosynthesis, and improves plant resistance to stresses.
- YaraMilaTM CEREAL fertilizer: Contains highly soluble and balanced proportions of macro- and micro-nutrients (23% nitrogen, 10% phosphate, 5% potassium oxide, 3% sulfur, 2% magnesium oxide, and 0.3% zinc). This fertilizer also contains sulfur and other elements that boost plant growth. Because the nitrogen is readily available, it does not increase the acidity of the soil. It releases phosphate over an extended period within a season, thus meeting the plant's needs throughout its growth. Because it has a lower phosphate concentration than DAP, Minjingu, and NPS, bulky amounts of YaraMilaTM fertilizer are required to meet recommended phosphate application rates for cereal production in most soils in ESA.
- Urea with 46% nitrogen: This has been used for many years by farmers, especially as a top-dressing for maize and other cereals, since the nitrogen stimulates strong plant growth.
- Calcium ammonium nitrate (CAN) with 26% nitrogen: Is used for top-dressing cereal crops and also supplies calcium and magnesium. The need for fertilizer is different depending on the location. Some farmers find it better to use fertilizer in areas of poor yields, where the need is greatest (target application). It should be noted that most legumes may not require the application of nitrogen, since the plants fix this nutrient through biological processes. Under severely depleted soil conditions, legumes may need a small amount of 'starter' nitrogen to support good crop development.

Benefits of the technology: Africa RISING research shows that adding the right amount of fertilizer results in more than double the maize yield when compared with farmer practice (continuous cultivation with little or no fertilizer). Under the environmental conditions of the highlands and medium-elevation zones of northern Tanzania, the trials showed that applying fertilizer achieves an increase of 120% or greater. Even under semi-arid environments, responses of about 57% are obtained. When moisture is inadequate, DAP and YaraMila CEREAL fertilizers are better sources of phosphate than the now replaced Minjingu Mazao fertilizer because it had low phosphate rock solubility (Savini et al., 2015), which could lead to a 10% yield reduction compared with DAP.

Table: *Benefits of fertilizer use on maize productivity*

Nutrient application	Agroecosystem	Average rainfall (mm/year)	Increase in productivity (Grain yield, %)	Gross margin (US$/ha)
P (20 kg/ha) + N (60 kg/ha)	Highlands of northern Tanzania	845	210 (153–444)	685–1,245
	Medium altitudes of northern Tanzania	743	215 (74–470)	820–1,215
	Lowlands of northern Tanzania	769	120 (63–300)	740–1,435
P (15 kg/ha) + N (60 kg/ha)	Semi-arid central Tanzania	408	57 (22–84)	102–357
P (18 kg/ha) + N (69 kg/ha)	Semi-arid central Malawi	855	128 (30–290)	n/a[1]
P (18 kg/ha) + N (69 kg/ha)	Semi-arid southern Malawi	723	187 (50–300)	263% (net income)

Notes: for northern Tanzania, work was undertaken in Babati district using DAP and Minjingu Mazao fertilizers. For semi-arid central Tanzania, work was done in districts of Kongwa and Kiteto using DAP, Minjingu, and YaraMila CEREAL fertilizers. Both YaraMilaTM and Minjingu supply additional nutrients needed by plants in small doses (micro-nutrients) and reduce soil acidity due to the presence of basic cations in their formulation. In Malawi, the work was conducted in Machinga and Dedza districts using NPS fertilizer and urea. 1 Trials across the semi-arid agroecosystems in Malawi showed that the right amount of fertilizer to apply is less than the highest recommended rate for maize of 92 kg nitrogen/ha. Applying 69 kg nitrogen/ha and 18 kg phosphate/ha increased maize yield by between 50% and 300% compared with systems using no fertilizer.

Actual maize yield varied quite widely depending on the use of a local maize type or improved variety, and on the use of fertilizer or farmer practice (little or no fertilizer). This is due to the combined effects of variable soil fertility, topography, and/or different agronomic practices adopted by farmers, such as planting density, time of planting, and weed management. The research shows that it is important to use fertilizer and to grow an improved maize variety with good agronomic practices, such as timely planting and proper plant spacing, to reduce variability in productivity. The increase in yield due to fertilizer application is sufficient, in most cases, to offset the cost of inputs and other operations, as noted by the marginal rate of return of at least 1.6 in Babati, Kongwa, and Kiteto districts, Tanzania.

Table: *Marginal rates of return on fertilizer application in Tanzania*

	Babati	Kongwa and Kiteto		
Fertilizer source	Low rainfall	High rainfall	Low rainfall	High rainfall
DAP	2.3	2.4	n/a[1]	n/a
Minjingu Mazao	2.1	1.6	1.9	2.3

Source: Adapted from Kihara et al. (2017; CC BY-SA 4.0). No data for DAP in Kongwa and Kiteto.

The results of the experiments show that farmers should apply fertilizers to decrease the chance of net financial loss. Without fertilizers, the chance of loss or of doubling invested capital are about the same, but when farmers apply fertilizers, the chance of loss is completely reduced and/or eliminated.

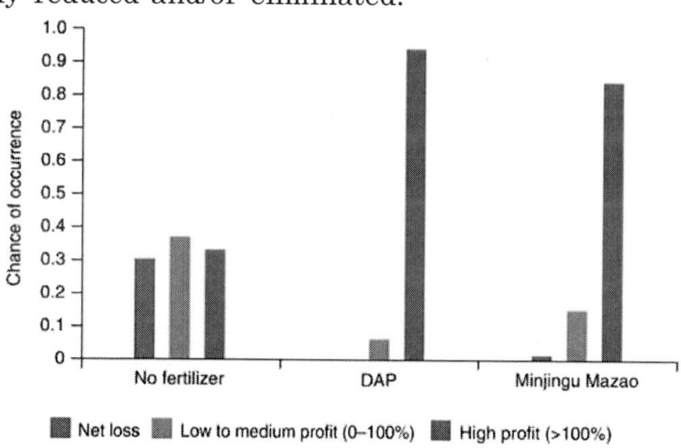

Figure: *Profit levels and chance of occurrence with and without fertilizers in Babati, Tanzania. On the y axis, 1 represents 100% chance of occurrence.*

For optimum benefit from these technologies, farmers should use improved agronomic practices, such as row planting at the appropriate spacing, weeding at the right time, and disease and pest control. Even without fertilizer, improved practices enhance crop productivity by 20-45%, and yield benefits increase further when improved practices are combined with fertilizer application (Figure 4.4).

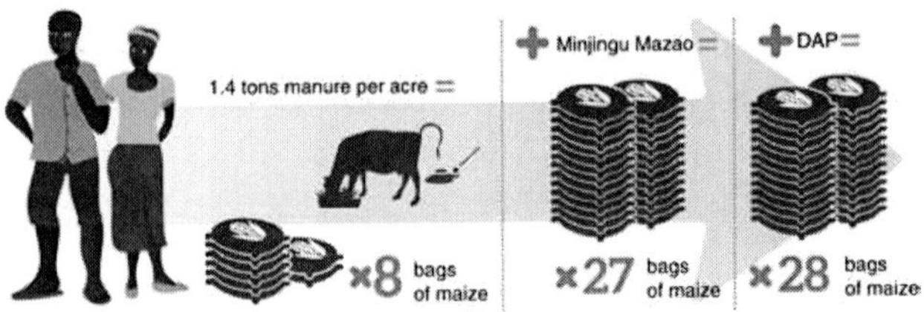

Figure: *Yield (on acre basis) obtained under farmer practice and with different fertilizers in Babati, northern Tanzania in the 2015/16 cropping season. Data are derived from participatory evaluation. One bag is equivalent to 100 kg.*

Farmers' responses: A clear demonstration of the benefits of fertilizer use with good agronomic practices, derived from farmer participatory evaluation (150 male farmers and 106 female farmers) was captured for the 2015/16 cropping season. Both female and male farmers perceived that the new technologies increase grain yield and net returns (income). In addition, male farmers perceived that the technologies reduce production risk. The farmers understood that using fertilizers will provide higher financial returns and reduce risk, as compared with their existing practice. However, the technology will also increase the requirement for cash, and those without access to finance may be unable to adopt it. Affordability is more important for women farmers than men; this could be due to women's limited access to financial resources. Providing loans for inputs is likely to enhance the adoption of this technology, particularly among women farmers.

Table: *Farmer qualitative assessment of industrial fertilizers in Babati, Tanzania*

Evaluation criteria	DAP	Minjingu Mazao	Male	Female	Male	Female
Accessibility	3.6	3.6	5.8	5.6		
Affordability	4.8	2.6	2.4	3.2		
Labor requirement	3.4	2.2	2.6	2.8		
Cash requirement	3.4	2.6	2.4	2.6		
Grain yield	7.0	7.4	7.0	7.2		
Net return	7.2	7.0	6.8	7.2		
Risk reduction	6.6	3.8	6.6	4.2		

Note: Rating score: 1 = the lowest and 10 = the highest. Figures greater than 5 are perceived to show an improved performance of the technology compared with the traditional practice; figures lower than 5 show a reduced performance.

Making the most of industrial fertilizer application: Industrial fertilizers undergo various changes in the soil, and these influence the solubility and availability of nutrients to plants. When plants receive nutrients through their roots, fertilizers should be added to the soil in a form that maximizes take-up of nutrients by the roots to support strong plant growth and development. Management practices for fertilizer application are guided by the four 'R's or 'rights': the right nutrient source (type of fertilizer) for the targeted crop(s); the right rate or quantity being applied; the right time of application; and the right place of application.

Figure: *The 4 'R's of nutrient stewardship.*

Right nutrient source: This is required to improve the agronomic performance of crops. Apply only the nutrients that are needed, based on a soil test. Applying a fertilizer with nutrients that are not needed decreases the efficiency and profitability of fertilizer use. Straight fertilizers, compound fertilizers, and specialty blends of micronutrients are available.

Soil pH and drainage conditions are also important factors to consider when selecting the right fertilizer.Industrial fertilizers are supplied as granules or liquids. Liquid fertilizers are for foliar application and are used mostly for specialty horticultural crops, while those supplied in solid form are used for soil application. Slow-release nitrogen fertilizers have become available more recently. Specialty and slow-release fertilizers are more expensive, but result in higher yields by ensuring the gradual release of small amounts of nutrients to the plant; they also reduce the need for top-dressing.

When soils are degraded and low in organic matter, it is advisable to add organic nutrients such as farmyard manure and compost, to improve the physical structure and increase the ability of the soil to hold water and nutrients. In addition, legumes grown in one season are a source of nitrogen for the next non-legume crop in a rotation or intercrop system, helping farmers save money on chemical nitrogen fertilizers. Right rate: The amount of fertilizer applied depends on the soil condition (i.e., level of limitation), rainfall, and affordability. Over-application reduces agronomic efficiency, lowers incomes, and contributes to environmental degradation (e.g., water pollution). Under-application results in inefficiency and a poor harvest. Hybrid crop varieties use fertilizers more efficiently than local varieties and, although they require larger amounts of fertilizer, application is likely to be profitable.

Application of 1 kg of nitrogen/ha resulted in 26 kg of hybrid maize grain yield, compared with only 17 kg of grain yield for local varieties using the same agronomic practices.Right time: Most nutrients and micronutrients are applied when the crop is first planted, to support the growth of the plants. Nutrients that move quickly through the soil (e.g., the nitrogen supplied by industrial fertilizers) need to be applied two or three times, usually one third at planting followed by top-dressing, with one third at four weeks, and one third at eight weeks after planting, or one third at planting and the remainder at about four weeks after planting.

Figure: *A farmer applying industrial fertilizer (urea top-dressing)*

The top-dressing must be applied when the soil is moist to ensure the nitrogen passes quickly into the soil. Under dry conditions, the nitrogen should be applied and then immediately covered with soil to avoid nitrogen being lost via volatilization. Slow-release fertilizer may be applied once at planting because the nitrogen is released gradually while the plant grows.

Basic requirements of good agricultural practice:

- Soil analysis is important to guide specific nutrient applications, especially for sites where this has not already been done. Analysis can be done by government or commercial laboratories, or by using portable laboratories, which are often cheaper.
- Use good-quality seed of improved crop varieties that are suitable for the area.
- Protect farms from surface runoff, since this carries away the topsoil and applied fertilizers.
- Create planting rows across the line of slope to avoid creating surface water runoff channels.
- Prepare the land early and plant on time. Late planting can reduce available seasonal water for optimum plant growth, and periods of water stress may coincide with critical crop growth stages.
- Plant seeds at an even depth of 2-5 cm (for maize and pigeonpea) in firm, moist soil to ensure good seed-to-soil contact. This maximizes moisture uptake and subsequent germination.
- Use the correct planting density and row spacing for each crop to maximize yields. Recommendations should be available from the seed provider or local extension agent.
- Keep the crop free from weeds, diseases, and pests. Fertilizer application at planting
- The choice of application method depends on the nature of the fertilizer, with the objective being to maximize nutrient uptake by the crops. The right amount of fertilizer should be added, based on analysis of limiting nutrients.
- Application in a planting hole: place fertilizer approximately 5 cm below or to the side of the seed. Do not place seeds in contact with the fertilizer as this can affect germination. For large grain crops such as maize, place fertilizer and seed at least half a finger deep and one finger apart.
- Application in bands: when plants are close together (e.g., beans, soybean), place the fertilizer in a line along the planting furrow. Mix the fertilizer with the soil before planting the seeds. The fertilizer depth depends on the crops and system,

as well as the type of fertilizer. Place immobile nutrient fertilizers (phosphates) at 10-15 cm below the soil surface.

- o Broadcasting: Spread the fertilizer over the soil surface and incorporate it in the soil through primary or light tillage. Do not use this method for large-seeded cereals and legumes.

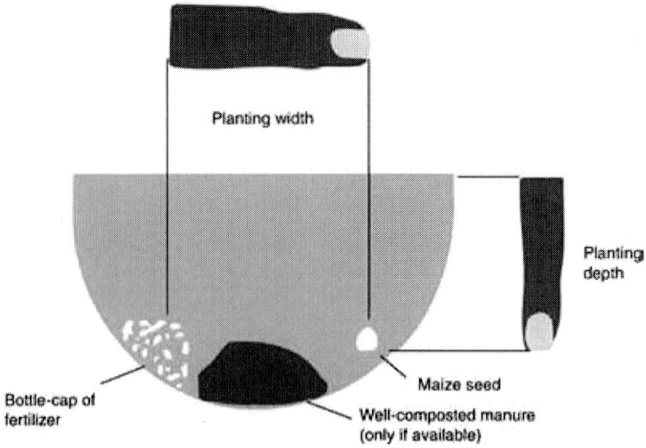

Figure: *Fertilizer application in a planting hole.*

At top-dressing

- o Use urea or CAN for top-dressing cereal crops.
- o For maize, apply at about four to six weeks after planting (or when the maize is knee high) in a weed-free field.
- o Do not top-dress legume crops in sole crops or intercrops.
- o Top-dress when soils are sufficiently moist; place the fertilizer next to the planting hill.
- o Cover the fertilizer with soil to minimize loss.
- o Weed at least twice and avoid weeding when the soil is dry, since excessive evaporation of soil moisture can cause plants to wilt.

Sustainability

- o Train farmers on the principles and procedures of using fertilizers properly, including the 4 'R's, the need for soil analysis, and where to get assistance.
- o Educate farmers on the options available to correct different soil nutrient deficiencies.
- o Train farmers on accompanying technologies needed to maximize the benefits of using fertilizers.

Description of the technology: Organic sources of nutrients are particularly important in degraded soils, where fertilizer uptake by plants may be limited by the soil's physical and biological properties. Farmyard manure is rich in nutrients, but poor methods of collection, storage, and application can lead to large losses of nutrients. Depending on the condition of the land, the quality of livestock feed, and application rate, farmyard manure can contain similar amounts of nitrogen, phosphorus, and potassium as industrial fertilizers, with nutrients released gradually through the season as the organic matter decomposes.

Manures and composts are accepted as 'right source' nutrients. The highest concentration of nutrients is provided by manure derived from poultry droppings, followed by goats and sheep, and then cattle manure. For each manure type, the nutrient concentration varies widely from place to place, and from farmer to farmer, depending on the method of collection, storage, materials fed to the livestock, and the composting process. Low-quality manure has a nitrogen content of less than 0.6%, while high-quality manure contains at least 1.6% nitrogen. Most chicken and goat or sheep manures are classified as high quality. Cattle fed on rations containing leguminous plants also produce high-quality manure, while cattle feeding mostly on cereals, especially free-ranging cattle, produce low-quality manure.

Table: *Indicative means and ranges of manure quality parameters for East and Southern Africa*

Manure	Nitrogen (%)	Phosphorus (%)	Potassium (%)	Carbon (g/kg soil)	Carbon/ nitrogen ratio
Cattle manure[1]	1.4 (0.5–2)	0.6 (0.2–1.6)	1.3 (0.5–2.7)	35 (17–52)	26 (17–56)
Manure/ compost[2]	1.1 (0.3–1.9)	0.3 (0.1–0.8)	2.4 (0.4–7)	24	23
Farmyard manure[3]	1.6	0.5	1.43	69.1[3]	7.5[4]
Goats/sheep[1]	1.5 (0.9–2.3)	0.4 (0.2–0.8)	1.4 (0.7–2.6)	32 (20–46)	22 (15–26)
Poultry[1]	1.7 (0.9–2.6)	1.3 (0.3–1.8)	0.7 (0.5–0.9)	35 (28–48)	20 (16–33)

Table: *Common convention on manure quality*

Category	Nitrogen content	Carbon:nitrogen ratio
High quality	>1.6%	<10
Low quality	<0.6%	>17

Source: Bationo et al. (2004).

Compost is made from the stalks and leaves of crops, grasses, and other vegetative matter that is not fed to livestock. The components (mixed materials such as crop residues, kitchen waste, animal droppings, and ash) are given time to break down to make a fine soil-like organic matter.

The materials are arranged in successive layers, either in a pit or on the soil surface, and kept moist under shade. The pile should be turned every three weeks or so and kept for at least three months until the composting process is complete. When compost is made mostly from legume residues, animal dung, and ash, it will be of high quality, while compost made mostly from cereal residues (e.g., maize, rice, and wheat), will be of low quality.

Compost and manures also contain a large amount and diversity of helpful microorganisms, which convert nutrients into forms that are easily accessible to plants. However, compost and manures are not usually available in sufficient quantities for application at the recommended rates of 10-15 t/ha, and the nutrients contained may be imbalanced, requiring combination or integration with industrial fertilizers. As with industrial fertilizers, use of organic resources and combining them with industrial fertilizers should follow the 4 'R's approach.

Benefits of the technology: Research conducted in Babati district, Tanzania in the 2014 and 2016 seasons showed that applying farmyard manure (6 t/ha) increased yields of maize and pigeonpea compared with the controls. Combining small amounts of industrial fertilizers with farmyard manure tripled the yields of both maize and pigeonpea. Applying farmyard manure also increased farmers' net incomes per hectare from US$ 434 to 865 relative to the control. Combining half the amount of farmyard manure with small amounts of Minjingu Mazao increased incomes still further, from US$ 865 to 1,647.

Africa RISING demonstration sites in the semi-arid zones of central Tanzania showed a mean maize grain yield of 2.9 t/ha when manure was applied alone at 5 t/ha, compared with a yield of 1.2 t/ha for farmers' traditional practice (with no or only small quantities of manure and no fertilizers).

This represents an increase of 145%. The mean maize grain yield was 3.4 t/ha when manure was supplemented with industrial fertilizer, an increase of 189% over farmers' traditional practice. Incomes increased accordingly. The maize grain yields and gross margins differed widely between farmer-managed plots, suggesting that farmers may obtain variable results for a given technology, depending on how they implement or adapt the technology to local conditions.

Table: *Changes in maize and pigeonpea grain yields and net incomes in Babati district, Tanzania, using farmyard manure*

Manure technology	Agroecology	Increase in productivity (Grain yield, %)	Gross margin (US$/ha)
Farmyard manure (6 t/ha)	Highlands of northern Tanzania	65 (41–93)	185–388
Farmyard manure (6 t/ha)	Medium altitudes of northern Tanzania	46 (0–139)	−12 to 162
Farmyard manure (6 t/ha)	Lowlands of northern Tanzania	No change	232–495
Farmyard manure + fertilizer[1]	Highlands of northern Tanzania	173 (123–283)	639–947
Farmyard manure + fertilizer[1]	Medium altitudes of northern Tanzania	141 (33–181)	542–799
Farmyard manure + fertilizer[1]	Lowlands of northern Tanzania	96 (24–199)	592–860
Farmyard manure (5 t/ha)	Semi-arid central Tanzania	145 (−19 to 304)	−235 to 955
Farmyard manure + fertilizer[2]	Semi-arid central Tanzania	189 (−21 to 425)	−294 to 1347

Additional benefits of the combined application of farmyard manure and fertilizer have been identified elsewhere. Residual benefits (i.e., yield increases beyond the season of application) are often realized and, with a long duration of application, benefits can persist for up to eight years. Fields applied with about 3.5 t/ha of manure for at least two of the previous five seasons produced an additional 800 kg/ha of maize compared with fields that had manure applied less frequently. Other known benefits of manure and compost include improved soil structure, aeration, and moisture-holding capacity, and these translate into the improved yields and profitability.

Although application of farmyard manure can increase yields, it is associated with some risk of net income loss due to the high labour costs of acquiring it and applying it to the fields. In some cases, this means the increase in yield may not be sufficiently great to offset the costs. Combining manure with industrial fertilizers eliminates the risk of loss and increases the chance of doubling the capital invested.

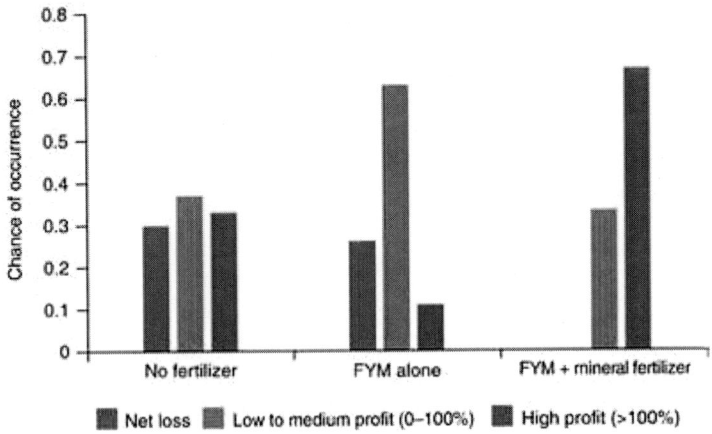

Figure: *Profit levels and their chance of occurrence with and without farmyard manure (FYM) in Babati, Tanzania. In the y axis, 1 represents 100% chance of occurrence.*

Farmers' responses: Manure is a locally available fertilizer resource. Many farmers use their own manure and attach no monetary value to the cost of labour involved in applying it to their fields. They often collect manure out of the cropping season when they are less busy and may apply it before tilling the lands. Experience shows that it may be better to apply the manure at the time of planting. It is important to ensure farmers know how to make good farmyard manure/ compost to maximize on yield benefits and to save cash by compensating industrial fertilizers with manures.

Making the most of organic fertilizer application: Right nutrient source: Compost and manures (particularly cattle manure) generally release nutrients into the soil more slowly than industrial fertilizers. They are therefore suitable for long-duration crops such as maize. Chicken manure releases nutrients more quickly and is therefore more suited to high-value, fast-growing crops such as vegetables.

Right rate: The quantity of compost and manure applied is normally based on the soil nutrient status, plant nutrient requirements, and the type and variety of crop, as well as the quality and availability of the manure and compost. The rate of application also depends on how often the manure is applied. Fields with a history of frequent application (every season) require less than fields that have had nothing applied for several seasons.

Right time: Because they release nutrients slowly, compost and manures are applied at the beginning of the crop cycle, at planting, or during or just before the land preparation stage. This choice depends on labour availability, volume, and quality of manure, and desired application method. Low-quality manure should be applied before planting to avoid competition for labour, especially when the planting window is short. High-quality manure should be applied immediately before sowing to prevent the nutrients being lost through leaching.

Right place: Manure of low quality can be applied by spreading followed by ploughing it under before the sowing time, banding (i.e., placing it in a line along a planting furrow), or putting it in the planting hole at planting time. With manure of high quality, e.g., poultry and goat or sheep manure, placement in the planting holes is the most appropriate solution, and especially for such crops as maize and pigeonpea, which are planted at a wide spacing (>25 cm apart). For other crops where spacing is small, banding is the most appropriate method.

Effective weed control: Timely planting allows the crop to establish early, use residual soil moisture, and outcompete weeds. Weeds compete with crops for nutrients, water, and light. The number of weeding cycles will depend on the density of weeds likely to cause crop losses. Generally, farmers should remove weeds at least twice to ensure good crop performance. Some weeds such as striga are parasitic and fertilizers can help to reduce the effects of these weeds. Good soil fertility management also promotes plant vigor, helping crops to outcompete and sometimes even smother weeds. Weeding should be avoided if the soils are dry, since the weeds can help to protect the soil surface and prevent excessive water evaporation.

Access to knowledge: Local extension officers, agro-dealers, and seed companies can advise farmers on which varieties of maize (or other crops) are best suited to a particular location. They can also advise on the most effective way to make good-quality compost, and which industrial fertilizers are best to complement application of farmyard manure. Look out for field days, agricultural shows, and other events hosted by research institutions, seed and fertilizer companies, and other non-governmental organizations.

How to get started: oEnsure there is plenty of well-decomposed farmyard manure or compost available in advance of planting the crop.

- Never apply partially rotten farmyard manure as it may scorch young seedlings. Apply manure and compost in the right place.

For large-seeded crops, apply manure in the planting hole and broadcast before planting (following the 4 'R's).

- Compost and manure must be incorporated in the soil and not left exposed to the sun or rain, since nutrients such as nitrogen will be lost to the atmosphere or leached into the soil. When applying to planting hills, bury the manure under 5 cm of soil.
- The quality of the manure or compost will determine the quantity to apply. This depends on the materials used to make the compost. Farmers can get advice from a local extension agent or, if possible, conduct chemical analysis. The quantity applied also depends on the amount of added industrial fertilizers (if any) and the nutrient demands of the target crop.
- When applying in hills, an appropriate container should be used to scoop only the right amount of manure per hill or hole. The container could be made from an old water bottle, for example.
- Some portions of the field will be more fertile than others. Target the manure or compost to the portions of the field that require it most (i.e., where yields are lower).
- It is better to apply moderate amounts of manure (about 3.5 t/ha) every season rather than larger quantities several years apart.
- Adopt good agricultural practices to derive the maximum benefits from the manure application.

The Economics of Fertilizers

The practical goal is to determine how much nutrient material to add. Since the farmer wants to know how much profit to expect if he buys extra fertilizer, the tests are interpreted as an estimation of increased crop production that will result from nutrient additions. Cost of nutrients must be balanced against value of crop or even against alternative procedures, such as investing the money in something else with a greater potential return. The law of diminishing returns is well exemplified in fertilizer technology. Past a certain point, equal inputs of chemicals produce less and less yield increase. The goal of the farmer is to use fertilizer in such a way that the most profitable rate is employed.

Fertilizers can aid in making profitable changes in farming. Operators can reduce costs per unit of production and increase the margin of return over total cost by increasing rates of application of fertilizer on principal cash and feed crops. They are then in a position to invest in soil conservation and other improvements that are needed when shifting acreage from surplus crops to other uses.

Farm manure: Among sources of organic matter and plant nutrients, farm manure has been of major importance in past years. Manure is understood to mean the refuse from stables and barnyards, including both excreta and straw or other bedding material, while the term fertilizer refers to chemicals. Large amounts of manure are produced by livestock; such manure has value in maintaining and improving soil because of the plant nutrients, humus, and organic substances contained in it.

As manure must be managed carefully in order to derive the most benefit from it, some farmers may be unwilling to expend the necessary time and effort. Manure must be carefully stored to minimize loss of nutrients, particularly nitrogen. It must be applied to the right kind of crop at the proper time. Also, additional fertilizer may be needed, such as phosphoric oxide, in order to gain full value of the nitrogen and potash that are contained in manure.

Manure is fertilizer graded as approximately 0.5-0.25-0.5 (percentages of nitrogen, phosphoric oxide, and potash), with at least two-thirds of the nitrogen in slow-acting forms. Commercial fertilizer equivalent to one ton (900 kilograms) of average manure can be purchased at a fairly low price. Furthermore, the expense of applying 100 pounds (45 kilograms) of 10-5-10 fertilizer is much less than the cost of applying 20 times as much manure. On properly tilled soils, the returns from fertilizer usually will be greater than from an equivalent amount of manure.

The application of manure to a crop cannot be controlled as readily as can granulated fertilizer. In general, manure does not provide all the plant nutrients needed and fails to provide any that cannot be supplied by artificial fertilizers. Thus, there is a tendency to discount the value of manure as fertilizer. In underdeveloped countries, however, where artificial fertilizer may be costly or unavailable and where labour is relatively cheap, manure is attractive as a fertilizer.

The main benefits of manure are indirect. It supplies humus, which improves the soil's physical character by increasing its capacity to absorb and store water, by enhancement of aeration, and by favouring the activities of lower organisms. Manure incorporated into the topsoil will help prevent erosion from heavy rain and slow down evaporation of water from the surface. In effect, the value of manure as a mulching material may be greater than is its value as a source of essential plant nutrients.

Green manuring: In reasonably humid areas, the practice of green manuring can improve yield and soil qualities. A green-manure crop is grown and plowed under for its beneficial effects, although during its growth it may be grazed. These green crops are usually annuals, either grasses or legumes, whose roots bear nodule bacteria capable of fixing atmospheric nitrogen.

Among the advantages of green-manure crops are the addition of nitrogen to the soil, increase in general fertility level, reduction of erosion, improvement of physical condition, and reduction of nutrient loss from leaching. Disadvantages include the chance of not obtaining a satisfactory growth; the possibility that the cost of growing the manure crop may exceed the cost of applying commercial nitrogen; possible increases in disease, insects, and nematodes (parasitic worms); and possible exhaustion of soil moisture by the crop.

Green-manure crops are usually planted in the fall and turned under in the spring before the summer crop is sown. Their value as a source of nitrogen, particularly that of the legumes, is unquestioned for certain crops such as potatoes, cotton, and corn (maize); for other crops, such as peanuts (groundnuts; themselves legumes), the practice is questionable. Farmers are gradually turning away from growing green-manure crops except where the crop may also serve as winter cover for the land.

Compost, peat, and sludge: Compost, peat, and sludge are used in agriculture and gardening as soil amendments rather than as fertilizers, because they have a low content of plant nutrients. They may be incorporated into the soil or mulched on the surface. Heavy rates of application are common.

Compost, or synthetic manure, is basically a mass of rotted organic matter made from waste-plant residues. Addition of nitrogen during decomposition is usually advisable. The result is a crumbly material that when added to soil does not compete with the crop for nitrogen. When properly prepared, it is free of obnoxious odours. Composts commonly contain about 2 percent nitrogen, 0.5 to 1 percent phosphorus, and about 2 percent potassium; if phosphate and potash are added while composting, those values are higher. The nitrogen of compost becomes available slowly and never approaches that available from inorganic sources. This slow release of nitrogen reduces leaching and extends availability over the whole growing season. Composts are essentially fertilizers with low nutrient content, which explains why large amounts are applied. The maximum benefits of composts on soil structure (better aggregation, pore spacing, and water storage) and on crop yield usually occur after several years of use.

In practical farming, the use of composted plant residues must be compared to the use of fresh residues. More beneficial soil effects usually accrue with less labour by simply turning under fresh residues; also, since one-half the organic matter is lost in composting, fresh residues applied at the same rate will cover twice the area that composted residues would cover. In areas where commercial fertilizers are expensive, labour is cheap, and implements are simple, however, composting meets the needs and is a logical practice.

Peat, composed of prehistoric plant remains that have accumulated under airless conditions in bogs, is a widely used organic soil amendment. Peat moss, the remains of sphagnum plants, is probably its most common form; it contains less than 1 percent nitrogen, with phosphorus and potassium below 0.1 percent. It is highly acid, with pH between 3 and 4.5 (a pH value of 7 is neutral and one above 7 basic). Peat improves the water-storage capability of soils and gives better structure to fine soils. Heavy applications of peat is usually the practice. It is used mostly by specialty-crop producers and on lawns and gardens.

Sewage sludge is the solid material remaining from the treatment of sewage. Its value for soil improvement depends on the method used for treating the sewage. Activated sludge, which results from aerobic (oxygen) treatment, contains 5 to 6 percent nitrogen and 1 to 3.5 percent of phosphorus. After suitable processing, it is sold as fertilizer and soil amendment for use on lawns, parks, and golf courses. It is rarely used in farming.

Liming: Liming to reduce soil acidity is practiced extensively in humid areas where rainfall leaches calcium and magnesium from the soil, thus creating an acid condition. Calcium and magnesium are major plant nutrients supplied by liming materials. Ground limestone is widely used for this purpose; its active agent, calcium carbonate, reacts with the soil to reduce its acidity. The calcium is then available for plant use. The typical limestones, especially dolomitic, contain magnesium carbonate as well, thus also supplying magnesium to the plant.

Another liming material is basic slag, a by-product of steel manufacture; its active ingredient is calcium silicate. Marl and chalk are soft, impure forms of limestone and are sometimes used as liming materials, as are oyster shells. Calcium sulfate (gypsum) and calcium chloride, however, are unsuitable for liming, for, although their calcium is readily soluble, they leave behind a residue that is harmful.

Lime is applied by mixing it uniformly with the surface layer of the soil. It may be applied at any time of the year on land plowed for spring crops or winter grain or on permanent pasture. After application, plowing, disking, or harrowing will mix it with the soil. Such tillage is usually necessary, because calcium migrates slowly downward in most soils. Lime is usually applied by trucks specially equipped and owned by custom operators.

Methods of application: Fertilizers may be added to soil in solid, liquid, or gaseous forms, the choice depending on many factors. Generally, the farmer tries to obtain satisfactory yield at minimum cost in money and labour.

Manure can be applied as a liquid or a solid. When accumulated as a liquid from livestock areas, it may be stored in tanks until needed and then pumped into a distributing machine or into a sprinkler irrigation system. The method reduces labour, but the noxious odours are objectionable. The solid-manure spreader conveys the material to the field, shreds it, and spreads it uniformly over the land. The process can be carried out during convenient times, including winter, but rarely when the crop is growing.

Application of granulated or pelleted solid fertilizer has been aided by improved equipment design. Such devices, depending on design, can deposit fertilizer at the time of planting, side-dress a growing crop, or broadcast the material. Fertilizer attachments are available for most tractor-mounted planters and cultivators and for grain drills and some types of plows. They deposit fertilizer with the seed when planted, without damage to the seed, yet the nutrient is readily available during early growth. Placement of the fertilizer varies according to the types of crops; some crops require banding above the seed, while others are more successful when the fertilizer band is below the seed.

The use of liquid and ammonia fertilizers is growing, particularly of anhydrous ammonia, which is handled as a liquid under pressure but changes to gas when released to atmospheric pressure. Anhydrous ammonia, however, is highly corrosive, inflammable, and rather dangerous if not handled properly; thus, application equipment is quite specialized. Typically, the applicator is a chisel-shaped blade with a pipe mounted on its rear side to conduct the ammonia five to six inches (13 to 15 centimetres) below the surface. Pipes are fed from a pressure tank mounted above. Mixed liquid fertilizers containing nitrogen, phosphorus, and potassium may be applied directly to the surface-by field sprayers where close-growing crops are raised. Large areas can be covered rapidly by use of aircraft, which can distribute both liquid and dry fertilizer.

The future for fertilizers: Future trends in fertilizer technology may be predicted by extrapolating from current developments. Mixtures and materials with high percentages of plant nutrients will dominate the field. Better ways of providing nitrogen, the most expensive of the three major nutrients, will be forthcoming, including increased use of anhydrous ammonia, ammonium nitrate, and urea. Nonleachable nitrogen, for example, can be obtained through the urea-formaldehyde (ureaform) reaction, and ammonium metaphosphate offers a concentrated liquid product. Micronutrients, or trace elements, specific to particular geographical areas will come into increasing use, as will custom mixing and bulk selling of mixtures containing several nutrients based on reliable soil and plant data.

"Complete environment" seeding in which seed, fertilizer, and water are incorporated in a biodegradable (decomposable in the soil) tape may come into use; with the tape planted, no further fertilizer or water will be needed until growth is well established. Such techniques using biodegradable tapes have already been developed on a small scale for use by home gardeners. Finally, larger and more precise fertilizing machines will be developed and adopted.

Factors in Cropping

Cropping Systems

The kind and sequence of crops grown over a period of time on a given area of soil can be described as the cropping system. It may be a pattern of regular rotation of different crops or one of growing only one crop year after year on the same area.

Crop rotation: Early agricultural experiments showed the value of crop rotations that included a legume sod crop in the regular sequence. Such a system generally maintains productivity, aids in keeping soil structure favourable, and tends to reduce erosion. Alfalfa, sweet clover, red clover, and Ladino clover are considered effective for building up nitrogen. Some legumes, however, do not leave nitrogen behind in the soil because it is deposited as protein in the harvested seed; soybeans are an example. Turning under the top growth of a legume aids in adding nitrogen. Though yields of grains are higher when they are rotated with legumes, it is difficult to determine how much of the improvement depends on the nitrogen added by the legume and how much on improved soil structure or fewer insects and disease.

The determination of the best rotation depends upon whether the crops compete with each other (i.e., if growing one crop lowers the yield of its successor) or complement each other; and the output of one crop on a given acreage leads to increased output of the other. This desirable complementary relationship exists only when one crop or soil-management practice concurrent with it provides nutrient or conditions required by the other crop. In this circumstance, grasses and legumes may complement grains or row crops by furnishing nitrogen, controlling erosion and pests, and improving soil structure to such an extent that greater production is achieved. The reverse can also occur; in certain prairie soils, continuous growing of deep-rooted legumes depletes soil moisture, and subsequent forage yield is improved by frequent plowing of the sod and planting of corn. In high-rainfall or irrigated areas, forage stands deteriorate from winter killing, disease, or grazing, to a point where a year of grain in the rotation allows an improved stand of forage later. Fallow (idle) land is complementary to wheat and other small grains in subhumid areas such as the Great Plains of the United States; such rotation is quite beneficial to wheat yield. Complementary relationships between crops can be terminated by the application of the physical law of diminishing returns, however, and give way to competition.

Both long-range and short-range profits motivate the farmer as cropping systems are examined in relationship to soil erosion. Excessive loss of soil to streams, rivers, and reservoirs is unacceptable to public policy as well as economically damaging to the farmer, and crop rotations that promote erosion are minimized. Soil losses are least from fields in continuous sod and most from continuous row crops. If row crops are grown in rotation with sod, the erosive susceptibility of row crops is reduced over a period of time. Peanuts (groundnuts), potatoes, tobacco, cotton, sugar beets, and some vegetables, and similar row crops that require frequent cultivation (intertillage) and leave minimal post-harvest residue are most likely to permit serious erosion. Less erosive are row crops such as corn (maize), sugarcane, and grain sorghum, which require less cultivation and leave more residue. Small grains such as wheat, oats, barley, and rye usually permit less erosion than the row crops. Among sod crops, grasses or grass-legume mixtures are less erosive than pure stands of legumes such as alfalfa. Fortunately, cropping systems that tend to control soil erosion usually tend also to give better yields than systems that promote excessive erosion. This results from increased availability of water to the plants and increased amounts of nutrients, which in erosive systems are washed away and lost.

Monoculture: The practice of growing the same crop each year on a given acreage, monoculture, has not been generally successful in the past, because nonlegume crops usually exhaust the nitrogen in the soil, with a resulting reduction in yields; this is particularly true in humid regions. The advent of low-cost nitrogen fertilizers, however, has induced reconsideration of the possible advantages of monoculture. These advantages can best be discussed in terms of a hypothetical general farm where it may be desirable to produce several different kinds of crops: the question to be answered is whether monoculture can do better than rotational systems in producing these crops while still maintaining productivity.

Advantages of Monoculture

First, if different kinds of soil exist on the farm, a monoculture system may permit each crop to be grown on the soil best suited to it. Forage crops, for example, could be confined to steep land to minimize erosion; intertilled crops could be planted on the better soils with gentle slopes. Wet areas could be used continuously for crops not requiring early-spring field operations, while dry soils could be used for drought-resistant crops such as sorghums or winter small grains.

Second, the fertility level of the soil can be adjusted to fit one crop more precisely than it can be adjusted to fit all the crops in a rotation.

Third, where continuous cropping is practiced and perennial forage crops are used, regular reseedings are avoided. This is an advantage, because each seeding is accompanied by the possibility of failure. Fourth, systems based on monoculture usually offer greater flexibility in planning the system to meet year to year changes in the need for various crops. Part of the acreage can be shifted from one crop to another without upsetting the total farm cropping plan.

Disadvantages of monoculture: On the other hand, requirements for successful monoculture are more demanding of management skill than are sod-based rotations. The entire nitrogen need of nonlegume crops must be met by purchased fertilizers or by use of manure. Closer attention to soil erosion is necessary, except for perennial sod. Soil-structure problems can become severe where intertilled crops are grown continuously. In monoculture, the farmer is completely dependent on chemical insecticides, disease-resistant plant varieties, soil fumigation, and similar methods of controlling insects and diseases that are usually controlled by crop rotation.

Thus, the choices of cropping systems that maintain good productivity, minimize soil losses, and are in harmony with demand and desired business organization are not easily made. The growing use of systems analysis will undoubtedly aid in rational selection among the bewildering array of possibilities.

Crop protection: Crops are vulnerable to attack, damage, and competition. Insects, plant disease, nematodes, rodents, weeds, and air pollution are among the many enemies that can reduce crop yields and deny man the use of some of his farm-stored crops.

Insects, for example, can destroy a crop in a relatively short time. Control measures for many years have engaged the attention of farmer and scientist, yet full success has not been achieved, and the battle continues. The problem is further complicated by the fact that control measures not only kill unwanted insects, but also may harm honey bees as well as the parasites and predators that destroy insect pests.

At least 10,000 species of insects are classed as unwanted. Of these, several hundred species are particularly destructive and require some degree of control. They destroy food as well as the forage, pasture, and grain needed to produce livestock; and, in addition, they carry and transmit many diseases of plants and animals.

Chemical control of insects: Insecticides generally are effective, cheap, and safe if handled correctly; the good derived from them, however, can be partly offset by adverse effects. Chlorinated hydrocarbon insecticides such as DDT, for example, may leave residues toxic to beneficial insects, fish, and other wildlife; the insecticides may be found in meat and milk, or they may persist in the soil. Another problem is that some species of insects build up resistance to chlorinated hydrocarbon, organic phosphate, and carbamate insecticides. These disadvantages can be overcome only by persistent search for new and safer insecticides accompanied by wide use of nonchemical insect control.

A wide range of organophosphate and carbamate materials is now available. These can be applied to avoid most of the problems related to residues. Malathion and carbaryl, for example, are used to control insects in areas where persistent materials might appear later in meat or milk and can also be applied in areas where fish and wildlife might be affected. Those two chemicals offer a broad range of toxicity to insect pests. Unlike chlorinated hydrocarbons, they can be applied up to within a day or so of harvest without harm to many crops; they are dangerous, however, to those who apply them and must be handled with care.

Some insecticides are effective in very small amounts. This fact has stimulated development of ultralow-volume technology, where special equipment permits dispersal of low volumes of undiluted chemicals, which offers cost advantages as well as drastic reduction of the chemicals in the environment. For example, six to 16 ounces (170 to 450 grams) per acre of Malathion may be effective against grasshoppers, boll weevil, cereal-leaf beetle (Oulema melanopus), mosquitoes, and the beet leafhopper (Circulifer tenellus). Formulation of chemicals in granules rather than sprays offers some advantages in use and applications; among others, it reduces the amount needed and also lessens the chance of adverse effects on beneficial insects and wildlife.

Certain insects that attack cotton, vegetables, and forage crops may be controlled by chemicals absorbed by the plant. Called systemics, they are placed with the seed at planting time. The chemical is taken up by the plant, and insects die when they attempt to feed on the leaf or stem. Beneficial insects that do not feed on the plant remain unharmed.

Nonchemical control of insects: Light traps that give off radiation that attracts insects have been under test for many years. They have been somewhat successful in controlling the codling moth (Carpocapsa pomonella) and the tobacco hornworm (Protoparce sexta).

Use of reflective aluminum strips, placed like a mulch in vegetable fields, has reduced or prevented aphid attack and thus protected cucumbers, squash, and watermelons from mosaic diseases. This technique may supplant insecticides, which frequently do not kill aphids quickly enough to prevent crop losses from virus transmitted by them.

For stored products, heat or cold can control many insects that frequent such places. Also, changing the proportions of oxygen, nitrogen, and carbon dioxide in the storage atmosphere can provide control. It has been discovered that, if adult Indian-meal moths (Plodia interpunctella) were exposed to certain wavelengths of sound during the egg-laying period, their reproduction was reduced by 75 percent. The sound waves had a similar effect on flour beetles (Tribolium species). Further development is needed, but this method offers potential as a nonchemical control. Other types of physical energy can also kill insects. Light waves, high-frequency electric fields, high-intensity radio frequencies, and gamma radiation have been investigated; some offer promise.

Certain cultural practices can prevent or reduce insect crop damage. These include destruction of crop residues, deep plowing, crop rotation, use of fertilizers, strip-cropping, irrigation, and scheduled planting operations. Such practices are useful but cannot be relied upon entirely to eliminate severe infestations.

Biological controls: The question of using biological controls has always been of considerable public interest. The control agents include parasites, predators, diseases, protozoa, and nematodes that attack the insect pests. Biological controls cannot replace insecticides entirely, because nature provides for survival of both beneficial and destructive insects. Before the population of a parasite or predator can expand, a high population of the host species must also be present. Sometimes the control agents are far outnumbered by the pest insect. Parasites and predators have furnished good control of the Japanese beetle (Popillia japonica), European corn borer (Pyrausta nubilalis), alfalfa aphid (Therioaphis maculata), alfalfa weevil (Hypera postica), and several others.

Microbial agents can be used for control. There exist about 1,100 viruses, bacteria, fungi, protozoa, rickettsiae, and nematodes that parasitize insects. Many pathogens are specific to a particular insect but are harmless to man and domestic animals. It is a possibility that insect pathogens can be produced, packaged, distributed, and applied in much the same way as insecticides.

The ideal solution to insect-control problems is to plant crop varieties that are resistant to attack. The only difficulty is that such varieties are not universally available, and development entails a very long process.

Sterilization of male insects by gamma radiation and their release into a population of wild insects is a promising approach. It has proved successful in control of screwworms and fruit flies, replacing chemicals in some areas. Chemical attractants, which lure insects into contact with small amounts of insecticide or a sterilant, also offer much promise.

All aspects of insect control considered, it is possible that "integrated control," coordinated employment of more than one method, may be the answer. Combining resistant varieties with a systemic insecticide that leaves the parasites and predators unharmed, for example, has been successful in combatting the spotted alfalfa aphid in California. Preliminary reduction of heavy infestation by chemical spray combined with bait, followed by the sterile-insect technique, provides another example of integrated control. Use of sex attractant in light traps, plus special management of postharvest residues, has controlled the tobacco hornworm. Other examples might be cited, but the principal value of such control methods lies in using less insecticide and thus contributing to maintenance of a good environment.

Control of plant diseases and nematodes: Insects, of course, are not the only agents hazardous to crops. Plant diseases and the microscopic worms called nematodes have the potential of creating wholesale destruction of crops, especially those grown in regions of wide weather fluctuation. In fact, these plant pests sometimes limit the kinds and varieties of crops that can be grown. The damage they cause may sometimes be mistaken for that caused by unfavourable weather. Epidemics may destroy crops completely.

As with insects, control of plant diseases and nematodes covers a broad spectrum of measures: use of chemicals, resistant varieties, quarantine, forecasting and warning, cultural practices, heat treatment, and others. Furthermore, most plant virus diseases are transmitted by insect carriers, so control of insects is linked to control of disease.

Nematodes and plant disease can at times be controlled fairly well by crop rotation, deep plowing, and burning of stubble and debris that remain after harvest. Though burning destroys aboveground organisms and permits economical control by chemicals, it contributes to air pollution and destroys organic matter. In another technique, propane-gas flame is applied to living plants as well as stubble to kill disease spores. A virus disease of sugarcane is controlled by heating diseased cuttings in hot-air ovens. Stem rot disease of peanuts (groundnuts) can be controlled by plowing under dead plant debris or by planting the seed on a raised bed followed by application of a preemergence weed killer.

Successful control of epidemic plant disease may depend on prompt application of chemicals before the disease outbreak. Many governments operate plant-disease forecasting and warning services for farmers. The service is based primarily on analysis of temperature, rainfall, humidity, and dew-all factors that can create conditions favourable to disease outbreaks.

Weed control: Weed control is vital to agriculture, because weeds decrease yields, increase production costs, interfere with harvest, and lower product quality. Weeds also impede irrigation water-flow, interfere with pesticide application, and harbour disease organisms.

Early methods of weed control included mowing, flooding, cultivating, smothering, burning, and crop rotation.

Though these methods are still important, other means are perhaps more typical today, particularly the use of herbicide (plant-killing) chemicals. Another technique is to introduce insects that attack only the unwanted plant and destroy it while leaving the crop plants unharmed.

The inadequacy of the cultural, mechanical, and biological control systems, however, stimulated the rapid development of chemical usage since World War II. Herbicides have had an impact on crop production, changing many cultural and mechanical agricultural operations.

Herbicides are formulated as wettable powders, granular materials, emulsions, and solutions. Any of them may be applied as a spot treatment, broadcast, placed in bands, or put directly on a specific plant part. When formulated as solutions or emulsions, the chemical is mixed with water or oil.

Spraying is the most common method, permitting extremely small amounts to be applied uniformly because of dilution. Sprays can be accurately directed underneath growing plants, and calibration and rate control are easier with spray machines than with granular applicators. Granular formulations have advantages under some conditions, however. The use of herbicides must be integrated into the overall farm program because the optimum date and application rate depend on the crop stage, the weed stage, weather conditions, and other factors.

Careful use of herbicides in farm production lowers cost, resulting in a more economical product for the consumer. Herbicides cut the costs of raising cotton, for example, by reducing labour requirements for weed control up to 60 percent. Herbicides replace hand labour in growing crops, labour that is no longer available in developed nations at costs the farmer can afford. Machines for chemical application are widely available.

When used as directed, herbicides are generally safe, not only for the operator but also for wildlife and livestock. The greatest difficulty lies in accidental injury to crop plants resulting from drift and from residues in the soil, particularly if residues enter water courses.

The future of chemical pesticides and herbicides is under debate by those who manufacture, sell, and use them and by those who are concerned about environmental quality.

The value of an assured food and fibre supply at reasonable cost is undeniable, and chemicals contribute much toward this. These substances also cause undesirable effects upon the environment, however, and indeed can be toxic to a wide range of organisms. This fact will demand an increasing amount of care in using chemicals, perhaps enforced by law, along with increasing use of nonchemical control techniques.

Harvesting and Crop Processing

Harvesting Machinery

Harvesting machinery is generally classified by crop: reapers for cutting cereal grains and threshers for separating the seed from the plant. The more modern combine cuts, threshes, and cleans the grain in one operation. Corn (maize) harvesting is performed by mechanical corn pickers that snap the ears from the stalk so that only the grain and cobs are harvested.

Corn shelling may be done mechanically in the field, after or with picking. Stripper-type cotton harvesters, which strip the entire plant of both open and unopened bolls, work best late in the season after frost has killed the green vegetative growth. Hay and forage machines include mowers, crushers, windrowers, field choppers, balers, and some machines that press the hay into wafers or pellets.

Grass, legumes, corn (maize), and other crops are often put into silos to keep them in a succulent and fermented state rather than stored dry as hay. To make silage, the crops must be cut up to permit tight packing in the silo, producing anaerobic fermentation and preventing formation of mold. Almost all silage crops are cut in the field with a forage harvester that cuts and chops the crop immediately or picks up and chops a windrow that has been cut and raked earlier.

Root crops are harvested with diggers and digger-pickers, which often pull up clods, stones, and vines with the crop. Though some machines carry workers who manually sort out extraneous material, this task is increasingly being performed mechanically. Modern sugar-beet harvesters lift the whole root from the ground, clean the earth from it, and deliver it to a bin or wagon. Sometimes the beet tops are removed before harvest of the roots and used for cattle feed. Peanuts (groundnuts) are lifted, vines and all, and allowed to dry before removal of the pods.

Tobacco-harvesting aids may be classified in three principal ways, according to the harvesting and curing methods used, which depend on the type of tobacco and its use. Flue-cured tobacco, a large plant that may stand three to four feet (90 to 120 centimetres) high, is harvested with machines that carry several workers who ride the lower platforms of the machines, cut the leaves, and place them on conveyor belts, where the leaves are tied mechanically or by hand. Burley tobacco has usually been harvested by workers using a machete-type knife. After cutting, the large end of the stalk is fixed onto the sharpened end of a stick, which-when loaded with a number of stalks-is hung by hand in a tobacco barn for curing. Researchers are attempting to mechanize the cutting, impaling, and hanging of burley tobacco. Little has been done, however, toward the mechanization of the harvesting of the small aromatic tobacco leaves, which are grown in the shade, picked by hand, tied with a string, then hung for curing.

Tree-crop harvesting is accomplished by hand or with mechanical shakers. Vegetable crops such as asparagus, lettuce, and cabbage are still harvested largely by hand, though scarcity and high cost of field labour has led to some mechanization in this area, notably with tomatoes.

Crop-processing machinery: Machinery is widely used to prepare crops for convenient transportation, for safe storage, for the market, and for feeding to livestock. Advances in such machines have been rapid, particularly with new crops, increased yields, multiple-crop practices, and changing techniques.

In the most common method of crop drying, the crop, usually grain, is spread on floors or mats and stirred frequently while exposed to the sun. Such systems, though extremely common in the underdeveloped countries, are very slow and dependent on the weather. Forced-air-drying systems allow the farmer much more freedom in choosing grain varieties and harvest time. Fairly simple in operation, these systems have been gaining popularity in the tropics. Heat is often added to increase air temperatures during the drying period.

In a process called dryeration, wet corn (maize) is placed in a batch or continuous dryer. After losing 10 to 12 percent of its moisture, the hot corn is transferred to the dryeration cooling bin, in which it is tempered for six to 10 hours and then slowly cooled by ventilation for 10 hours. This process reduces kernel damage and increases dryer output.

High moisture in stored hay not only causes rapid deterioration of its value as feed but often results in spontaneous combustion. When hay is first cut, it usually contains 70 percent or more moisture. It wilts and quickly dries to a moisture content of about 40 percent. At this stage, it can be dried to a safe storage condition, about 15 percent moisture, by blowing air through it, sometimes with supplemental heat.

Feed-processing mills, often referred to as feed grinders, are used principally for milling cereals for livestock feed, which aids digestion. The ground material is usually fairly coarse and at times may only be crushed. Modern mills frequently are designed to allow the farmer to grind the grain and to mix in various other ingredients in desired quantities.

Other types of crop-processing machinery include machines that separate desirable seed from weed seed, stems and leaves, and dirt; grading machinery to classify seed by width, length, or thickness; fruit graders and separators; and cotton gins, which separate cotton seeds from the fibres.

Regional Variations in Technique

Dryland Farming

Dryland farming refers to production of crops without irrigation in regions where annual precipitation is less than 20 inches (500 millimetres). Where rainfall is less than 15 inches (400 millimetres) per year, winter wheat is the most favoured crop, although spring wheat is planted in some areas where severe winter killing may occur. (Grain sorghum is another crop grown in these areas.) Where some summer rainfall occurs, dry beans are an important crop. All dryland crop yield is mainly dependent on precipitation, but practices of soil management exert great influence on moisture availability and nutrient supply.

Where rainfall exceeds 15 inches (380 millimetres), the variety of crop possibilities is increased. In areas of favourable soils and moisture, seed alfalfa is grown, as is barley. Some grass seed may be grown, particularly crested wheat grass of various types.

Fallow system and tillage techniques: Dryland farming is made possible mainly by the fallow system of farming, a practice dating from ancient times. Basically, the term fallow refers to land that is plowed and tilled but left unseeded during a growing season. The practice of alternating wheat and fallow assumes that by clean cultivation the moisture received during the fallow period is stored for use during the crop season. Available soil nitrogen increases and weeds are controlled during the fallow period. One risk lies in the exposure of soil while fallow, leaving it susceptible to wind and water erosion. Modern power machinery has tended to reduce this risk.

Procedures and kinds of tillage that are comparatively new have proved effective in controlling erosion and improving water intake. Moldboard and disk plows are being replaced with chisels, sweeps, and other tools that stir and loosen the soil but leave the straw on the surface. Where the amount of straw or residue remaining from the previous crop is not excessive, this trashy fallow system works well, and tillage implements are designed to increase its effectiveness.

Contour tillage helps to prevent excessive runoff on moderate slopes. Broad terraces can aid in such moisture conservation. Steeper slopes are planted to permanent cover.

Compacted zones at a depth of five to eight inches (13 to 20 centimetres) can be caused by tillage. As such zones interfere with storage of moisture, they can be controlled by growing deep-rooted alfalfa at intervals, or the compacted zone can be broken by fall tillage with chisels or sweeps set to a depth just below the zone of compaction. Such deep tillage will result in reduced runoff and deeper moisture penetration.

When using power machinery in dryland farming, the timing of operations is important. The soil is broken in the fall or early spring before weeds or volunteer grain can deplete the moisture. Use of a rod weeder or similar equipment during fallow can control the weeds. Planting is timed to occur during the short period in fall or spring when temperature and moisture are favourable.

Fertilizer use: Fertilizer is an important component of dryland technology. For example, 20 pounds per acre (22 kilograms per hectare) of nitrogen are recommended where rainfall is less than 13 inches (330 millimetres), ranging up to 60 pounds per acre (67 kilograms per hectare) where more rain is available; those figures refer to the production of wheat, but they are applicable to other dryland-farming areas. Where average annual precipitation is less than 12 inches (300 millimetres), the use of nitrogen is limited to years where moisture outlook is exceptionally favourable. Nitrogen fertilizer can be applied either in fall or spring. Band placement or broadcast techniques are utilized. Good results are obtained from broadcast spring application of nitrate fertilizer, and fall application of ammonia has also been successful. Local climates and rainfall patterns also determine choice of fertilizer and time of application.

Crops and Planting Methods

Figure: *Contour farming and strip cropping on sloping farmland.*

Alfalfa grown for seed on drylands is planted in rows, usually two to three feet (60 to 90 centimetres) apart; cultivation between rows is required during the first year. Alfalfa is also grown for forage where favourable. This practice builds nitrogen and organic matter, while improving soil structure. These legumes can be rotated with wheat if rain is between 16 and 18 inches (400 and 450 millimetres) and will increase the yield of wheat.

Other crops, such as cotton, peanuts (groundnuts), and grain sorghum, can be grown successfully in dryland agriculture. Where those crops are produced on sandy soils, special techniques are necessary to reduce soil blowing and drifting. Cotton and peanuts do not produce sufficient crop residues for protection from wind erosion, while sorghum does. For this reason, many farmers in such areas use various row combinations of cotton or peanuts with grain sorghum;

two rows of grain sorghum and four to eight rows of peanuts in alternating strips is a popular technique. Another is to use a two-year rotation of cotton and grain sorghum, in which two rows are cropped and two rows are fallow. These systems not only afford protection from wind erosion but also promote effective use of soil moisture.

Tropical farming: The area of the world bounded roughly on the north by the Tropic of Cancer and on the south by the Tropic of Capricorn, a vast land that embraces large parts of Latin America, Africa, India, Australia, and Southeast Asia, contains climates less favourable to agriculture and human settlement than those of the temperate zones. Within this Equator-centred area occur the climates known as tropical, which are characterized by two general types: warm and wet, and warm with partly deficient rainfall. In either, the total precipitation is usually quite heavy, which leaches the tropical soils of nutrients.

The area also has high temperatures with little variation the year round. The combination of high temperature and high rainfall causes organic matter to decompose quickly, leaving the soil deficient in humus. Vegetation flourishes in the tropics, along with weeds, insects, and disease organisms. Important climatic variations occur, depending upon land elevation.

Tropical crops include coconut, palm oil, rice, sugar, pineapple, sisal, cocoa, tea, coffee, jute, rubber, pepper, banana, and many others. In certain highland tropical areas, however, the crops common to temperate-climate agriculture can also be grown. The amount of tropical land well-suited to agriculture, however, is limited.

Plant-pest problems: The abundance of plant pests in the tropics, including weeds and disease, makes agriculture successful mainly in the plantation system, where needed control measures can be financed. The alternative is to move from deteriorated land to newer fields; this practice of shifting agriculture has also been common, because tropical soils lose their productive capacity so rapidly. The practice probably cannot be continued indefinitely, however, because of increasing population pressure.

The largest quantities of commercial tropical products originate in plantations, where skilled management is combined with sufficient capital to provide mechanized equipment. This is particularly true in the production of coffee, cocoa, rubber, coconut, banana, pineapple, sugarcane, and others. Much rice is produced in the Asian tropics and

Indonesia, however, on small farms with intensive hand labour and simple tools, where the prime mover is likely to be the ox or the water buffalo, not the tractor.

Water Management

Figure: *Know about the water management crisis in Australia including the development of a mathematical system to get a complete water balance*

Drainage, irrigation, and other special techniques of water management are important in tropical agriculture. An example is the cultivation of rice and sugarcane in the fertile coastal areas of Guyana. Originally through private enterprise and later by government efforts, large coastal areas were "empoldered" (diked) to keep back the sea in front and floods from the rivers in the rear. With a mean annual rainfall of 90 inches (2,300 millimetres), drainage is a critical factor; in fact, the system cannot discharge all possible floodwater, and so the crops must tolerate occasional drowning. With gravity drainage effective only at low tide, the drainage gates are opened on the ebbing tide and closed on the rising tide.

Great difficulty is encountered in keeping the outlets unclogged by the heavy sediment discharge. Since rain does not always fall when it is needed, many fields are irrigated. Most of the rice soil is specially tilled after plowing in order to create a better seedbed under the water, using tractors operating in water four to six inches (10 to 15 centimetres) deep. After this special tillage, the seeds are broadcast in one to two inches (2.5 to five centimetres) of water. Though maintenance and operation of such an intricate water-control system are not simple, Guyana rice production has been doubled through its use.

Mechanical problems: Mechanization faces many obstacles before wide adoption is possible in tropical regions. Difficult soils, stones, stumps, abundant labour, resistance from farmers, lack of incentives, lack of skills, lack of capital, low wages, high cost of machines, lack of dealer service, fragmented land ownership, all contribute to slow development of mechanization. Tropical soils differ markedly from those in the countries that manufacture land-preparation machinery, making adaptation of new design necessary.

The encountering of stones, wood, trash, and termite mounds causes machines to break down. Depressing climatic conditions reduce the performance of the machine operators. Tropical farm regions are notoriously irregular or mountainous, impeding intensive machine culture. The best soils in Brazil require special erosion controls, reducing the potential for large-scale mechanization. One of the greatest overall impediments to mechanization is the fear that unemployment might result from it, a failure to understand that economic development and higher living standards depend partly on increasing the productivity of labour.

As an example of the problems encountered in mechanizing tropical crops, the harvesting experience of a large sugarcane plantation in Trinidad is illuminating. On flatland of some 30,000 acres (12,000 hectares), the cane is grown on heavy clay soil in a climate with 50 inches (1,300 millimetres) of rain during the seven-month wet season and 10 inches during the five-month dry season. By 1960 the rising wage rate made harvest mechanization imperative. First, the traditional "bed" system, which functioned to remove floodwater, was changed to ridge planting; this made it possible for machines to operate and was a remarkable change in itself. Then it was decided to harvest with the cane combine, which tops, cuts, chops, and loads the chopped cane into transport vehicles. Although the combine is complicated and requires considerable power, it was deemed better than mechanical half-measures.

By 1969 the combines, however, were harvesting only 12.8 percent of the flatland crop, indicating that mechanization was far from complete. Three factors were responsible: first, the cane combines required extensive maintenance plus very expensive replacement parts. Second, it was difficult to mobilize a transport system to receive the output of the combines with any degree of economy. Third, the social problem of displaced workers had to be considered. The combines increased labour productivity sixfold over hand harvesting; thus, their introduction had to be slowed until surplus workers could be accommodated elsewhere. The limited success of this mechanization project indicates how complicated such a process really is.

Taking the largest view of possibilities for improving tropical agriculture, the most promising inputs of technology are improved crop varieties and increased use of fertilizers.

Hydroponics: The term hydroponics denotes soilless culture of plants. The possibilities of this technique have received considerable attention in recent years. In hydroponics, an outgrowth of laboratory techniques long used by scientists, plants are grown with their roots immersed in a water solution containing necessary minerals or rooted in a sand medium kept moistened by such a solution. Soilless culture of plants is similar in principle but larger in scale. A typical hydroponics technique has plants supported in a bed of peat, wood fibre, or similar material, on a wire screen with the roots dipping into the solution below. Aeration of the solution is provided. In another method, the plants are rooted in a medium of sand or gravel contained in a shallow tank into which the solution is pumped at intervals by automatic control. Between pumpings, the solution drains slowly down into a reservoir tank. Hydroponic techniques are practiced on a small scale both out-of-doors and in greenhouses.

Of the elements known to be necessary for plant growth, carbon, oxygen, and hydrogen are obtained by the plant from atmospheric gases or from soil water. The others are all obtained as mineral salts from the soil. The elements absorbed as salts-iron, manganese, boron, copper, zinc, and molybdenum-are required in minute quantities and are called the micronutrients. The principal elements that must be provided as dissolved salts in hydroponic techniques are nitrogen, phosphorus, sulfur, potassium, calcium, and magnesium. Numerous solutions have been devised to fulfill these requirements.

Crop yields of some plants can be obtained fully equal to those obtained on fertile soils. Wide-scale crop production by hydroponics, however, would be economic only for certain intensive types of agriculture or under special conditions. Some greenhouse crops, both vegetables and flowers, are grown by this method. In regions having no soil or extemely infertile soil but with favourable climate, hydroponic techniques have been very useful; for example, on some of the coral islands of the Pacific.

Greenhouses: The greenhouse is typically a structure whose roof and sides are transparent or translucent, permitting a sufficient quality and quantity of solar radiation to enter the structure for photosynthesis. It allows the growing of crops independently of the outside climate, since its interior temperature and humidity can be controlled. Greenhouses vary in size and complexity from small home or hobby structures to large commercial units covering an acre or more of land. An even smaller greenhouse might be termed the hot bed, a glass-topped box containing fermenting organic matter; the fermentation process yields heat, allowing the gardener to start plants from seed in early spring for later transplanting. The basic construction of a greenhouse consists of a light but sturdy frame capable of resisting winds and other loads. Conventional foundations usually support vertical walls; the roof may be gabled, trussed, or arched. The conventional greenhouse is fitted with glass panes, but plastic-film or fibre-glass panels often supplant glass.

Maintenance of temperature within the greenhouse is difficult because of fluctuating outside conditions. When the sun shines brightly, little heat is needed, and the heating system must be controlled in some way to prevent injury to the crop. Hot water, steam, electric cable, or warm-air furnaces provide the heat, which is usually controlled by thermostat. Temperatures in greenhouses are regulated to suit the crop. Typical ranges are from 40° F (4° C) for lettuce, violets, carnations, and sweet peas to 70° F (21° C) for cucumbers, tomatoes, and orchids.

Cooling is often required during summer days in warm climates. Ventilation is the simplest technique, reducing inside temperature to near that of the outdoors. Additional cooling by refrigeration may be required; in dry regions, the evaporative cooler is efficient and also increases the relative humidity within the structure. Another form of environmental control consists of adding extra carbon dioxide to the air if the crop requires it for extra photosynthetic efficiency.The commercial-greenhouse operator usually grows vegetables or ornamental plants. Such production makes more demands on the grower, because he must assume many of the tasks normally handled by nature in the open fields. He must regulate the temperature, ventilate, adjust the amount of entering sunlight, provide soil moisture, fertilize, and even facilitate pollination. During the off-season, the structure must be cleaned and fumigated, its soil restructured, and mechanical equipment checked. Mechanization of greenhouse operations has lagged far behind the pattern of agriculture in general. Disease is a particularly serious hazard in greenhouse farming, requiring constant attention and use of chemicals.

The Factor of Weather

Figure: *Know about the use of drones in agriculture for monitoring the environment of the field and enhance the quantity and quality crop*

The interaction of weather and living systems is a basic aspect of agriculture. Although great strides in technology have resulted in massive production increases and improved quality, weather remains an important limiting factor. Though man is not yet able to change the weather, except on a very small scale, he is capable of adjusting agricultural practices to fit the climate. Thus, weather information is of utmost importance when combined with other factors, such as knowledge of crop or livestock response to weather factors; the farmer's capability to act on alternative decisions based on available weather information; existence of two-way communication by which specific weather forecasts and allied information can be requested and distributed; and the climatic probability of occurrence of influential weather elements and the ability of the meteorologist to predict their occurrence.

Other weather-research benefits: Apart from the many applications of weather forecasting to current problems, meteorological research may benefit agriculture in at least three other ways: (1) improved planning of widescale land usage depends partly on detailed knowledge of plant-climate interactions; radiation, evapotranspiration, diurnal temperature range, water balance, and other parameters are measured and analyzed before a plan realizing maximum economic benefit for a given area is prepared; (2) agronomic experiments are combined with climatological documentation to obtain the greatest scientific and technological return; (3) problems of irrigation, row spacing, timing of fertilizer application, variety selection, and transplanting can best be solved with the aid of climatic environmental data; cultural practices related to artificial modification of microclimates should be based on research knowledge rather than personal judgment.

Nutrient Management in Problem Soils

Soil management is the application of operations, practices, and treatments to protect soil and enhance its performance (such as soil fertility or soil mechanics). It includes soil conservation, soil amendment, and optimal soil health. In agriculture, some amount of soil management is needed both in nonorganic and organic types to prevent agricultural land from becoming poorly productive over decades. Organic farming in particular emphasizes optimal soil management, because it uses soil health as the exclusive or nearly exclusive source of its fertilization and pest control. Soil management is an important tool for addressing climate change by increasing soil carbon and as well as addressing other major environmental issues associated with modern industrial agriculture practices. Project Drawdown highlights three major soil management practices as actionable steps for climate change mitigation: improved nutrient management, conservation agriculture (including No-till agriculture), and use of regenerative agriculture.

Environmental impact According to the EPA, agricultural soil management practices can lead to production and emission of nitrous oxide (N2O), a major greenhouse gas and air pollutant. Activities that can contribute to N2O emissions include fertilizer usage, irrigation and tillage. The management of soils accounts for over half of the emissions from the Agriculture sector. Cattle livestock account for one third of emissions, through methane emissions. Manure management and rice cultivation also produce emissions. Using biochar may decrease N2O emissions from soils by an average of 54%. the usage of artificial fertilizer in the agricultural field it leads to nutrition imbalance in the soil. Soils can sequester carbon dioxide (CO2) from the atmosphere, primarily by storing carbon as soil organic carbon (SOC) through the process of photosynthesis. CO2 can also be stored as inorganic carbon but this is less common. Converting natural land to agricultural land releases carbon back into the atmosphere. The amount of carbon a soil can sequester depends on the climate and current and historical land-use and management. Cropland has the potential to sequester 0.5-1.2 Pg C/year and grazing and pasture land could sequester 0.3-0.7 Pg C/year. Agricultural practices that sequester carbon can help mitigate climate change. Intensive farming deteriorates the functionality of soils. Methods that significantly enhance carbon sequestration in soil include no-till farming, residue mulching, cover cropping, and crop rotation, all of which are more widely used in organic farming than in conventional farming. Because only 5% of US farmland currently uses no-till and residue mulching, there is a large potential for carbon sequestration. Similar practices such as arable land conversion to grasslands, crop residues and cover crops have been proposed in Europe.

Practices: Conventional agriculture is driven by industrialization and aims to maximize efficiency. Practices include large-scale farming that specializes in monoculture and uses pesticides, herbicides, and fertilizers. Alternatives include conservation, regenerative, and organic agriculture, which can be broadly grouped as sustainable agriculture. Conservation agriculture has three main practices: minimizing soil disturbance, maintaining permanent soil coverage, and diversifying crop species. Similarly, regenerative agriculture practices use minimal to no tillage, cover crops, crop rotations, compost, and grazing. Organic agriculture incorporates most of these practices and emphasizes biological, not synthetic, management. There are three overarching practices that improve carbon sequestration in soils: increasing biomass inputs, decreasing SOC losses, and increasing the mean residence time (MRT) of SOC.

Figure: *Cover cropping and mulching practiced as soil management in Palau*

Specific soil management practices that affect soil health include:

- Controlling traffic on the soil surface helps to reduce soil compaction, which can reduce aeration and water infiltration.

- Planting cover crops that keep the soil anchored and covered in off-seasons so that the soil is not eroded by wind and rain.
- Crop rotations for row crops alternate high-residue crops with lower-residue crops to increase the amount of plant material left on the surface of the soil during the year to protect the soil from erosion.
- Nutrient management can help to improve the fertility of the soil and the amount of organic matter content, which improves soil structure and function.
- Tilling the soil, or tillage, is the breaking of soil, such as with a plough or harrow, to prepare the soil for new seeds. Tillage systems vary in intensity and disturbance. Conventional tillage is the most intense tillage system and disturbs the deepest level of soils. At least 30% of plant residue remains on the soil surface in conservation tillage. Reduced-tillage or no-till operations limit the amount of soil disturbance while cultivating a new crop, and help to maintain plant residues on the surface of the soil for erosion protection and water retention.
- Adding organic matter to the soil surface can increase carbon in the soil and the abundance and diversity of microbial organisms in the soil.
- Using fertilizers increases nutrients such as nitrogen, phosphorus, sulfur, and potassium in the soil. The use of fertilizers influences soil pH and often acidifies soils, with the exception of potassium fertilizer. Fertilizers can be organic or synthetic.
- Using a perennial grain crop such as Thinopyrum intermedium.

Figure: *Nitrogen fertilizer being applied to growing corn (maize) in a contoured, no-tilled field in Iowa.*

Nutrient management is the science and practice directed to link soil, crop, weather, and hydrologic factors with cultural, irrigation, and soil and water conservation practices to achieve optimal nutrient use efficiency, crop yields, crop quality, and economic returns, while reducing off-site transport of nutrients (fertilizer) that may impact the environment. It involves matching a specific field soil, climate, and crop management conditions to rate, source, timing, and place (commonly known as the 4R nutrient stewardship) of nutrient application.

Important factors that need to be considered when managing nutrients include (a) the application of nutrients considering the achievable optimum yields and, in some cases, crop quality; (b) the management, application, and timing of nutrients using a budget based on all sources and sinks active at the site; and (c) the management of soil, water, and crop to minimize the off-site transport of nutrients from nutrient leaching out of the root zone, surface runoff, and volatilization (or other gas exchanges).

There can be potential interactions because of differences in nutrient pathways and dynamics. For instance, practices that reduce the off-site surface transport of a given nutrient may increase the leaching losses of other nutrients. These complex dynamics present nutrient managers the difficult task of achieve the best balance for maximizing profit while contributing to the conservation of our biosphere.

Nutrient Management Plan

Figure: *Manure Spreader*

A crop nutrient management plan is a tool that farmers can use to increase the efficiency of all the nutrient sources a crop uses while reducing production and environmental risk, ultimately increasing profit. Increasingly, growers as well as agronomists use digital tools like SST or Agworld to create their nutrient management plan so they can capitalize on information gathered over a number of years. It is generally agreed that there are ten fundamental components of a crop nutrient management plan. Each component is critical to helping analyze each field and improve nutrient efficiency for the crops grown. These components include:

Field map: The map, including general reference points (such as streams, residences, wellheads etc.), number of acres, and soil types is the base for the rest of the plan.

Soil test: How much of each nutrient (N-P-K and other critical elements such as pH and organic matter) is in the soil profile- The soil test is a key component needed for developing the nutrient rate recommendation.

Crop sequence: Did the crop that grew in the field last year (and in many cases two or more years ago) fix nitrogen for use in the following years- Has long-term no-till increased organic matter- Did the end-of-season stalk test show a nutrient deficiency- These factors also need to be factored into the plan.

Estimated yield: Factors that affect yield are numerous and complex. A field's soils, drainage, insect, weed and crop disease pressure, rotation and many other factors differentiate one field from another. This is why using historic yields is important in developing yield estimates for next year. Accurate yield estimates can improve nutrient use efficiency.

Sources and forms: The sources and forms of available nutrients can vary from farm-to-farm and even field-to-field. For instance, manure fertility analysis, storage practices and other factors will need to be included in a nutrient management plan. Manure nutrient tests/ analysis are one way to determine the fertility of it. Nitrogen fixed from a previous year's legume crop and residual effects of manure also affects rate recommendations. Many other nutrient sources should also be factored into this plan.

Sensitive areas: What's out of the ordinary about a field's plan- Is it irrigated- Next to a stream or lake- Especially sandy in one area- Steep slope or low area- Manure applied in one area for generations due to proximity of dairy barn- Extremely productive-or unproductive- in a portion of the field-

Are there buffers that protect streams, drainage ditches, wellheads, and other water collection points- How far away are the neighbours-

What's the general wind direction- This is the place to note these and other special conditions that need to be considered.

Recommended rates: Here's the place where science, technology, and art meet. Given everything you've noted, what is the optimum rate of N, P, K, lime and any other nutrients- While science tells us that a crop has changing nutrient requirements during the growing season, a combination of technology and farmer's management skills assure nutrient availability at all stages of growth. No-till corn generally requires starter fertilizer to give the seedling a healthy start.

Recommended timing: When does the soil temperature drop below 50 degrees- Will a N stabilizer be used- What's the tillage practice- Strip-till corn and no-till often require different timing approaches than seed planted into a field that's been tilled once with a field cultivator. Will a starter fertilizer be used to give the seedling a healthy start- How many acres can be covered with available labour (custom or hired) and equipment- Does manure application in a farm depend on a custom applicator's schedule- What agreements have been worked out with neighbours for manure use on their fields- Is a neighbour hosting a special event- All these factors and more will likely figure into the recommended timing.

Recommended methods: Surface or injected- While injection is clearly preferred, there may be situations where injection is not feasible (i.e. pasture, grassland). Slope, rainfall patterns, soil type, crop rotation and many other factors determine which method is best for optimizing nutrient efficiency (availability and loss) in farms. The combination that's right in one field may differ in another field even with the same crop.

Annual review and update: Even the best managers are forced to deviate from their plans. What rate was actually applied- Where- Using which method- Did an unusually mild winter or wet spring reduce soil nitrate- Did a dry summer, disease, or some other unusual factor increase nutrient carryover- These and other factors should be noted as they occur.

When such a plan is designed for animal feeding operations (AFO), it may be termed a "manure management plan." In the United States, some regulatory agencies recommend or require that farms implement these plans in order to prevent water pollution. The U.S. Natural Resources Conservation Service (NRCS) has published guidance documents on preparing a comprehensive nutrient management plan (CNMP) for AFOs.

The International Plant Nutrition Institute has published a 4R plant nutrition manual for improving the management of plant nutrition. The manual outlines the scientific principles behind each of the four Rs or "rights" (right source of nutrient, right application rate, right time, right place) and discusses the adoption of 4R practices on the farm, approaches to nutrient management planning, and measurement of sustainability performance.

Nitrogen management: Of the 16 essential plant nutrients, nitrogen is usually the most difficult to manage in field crop systems. This is because the quantity of plant-available nitrogen can change rapidly in response to changes in soil water status. Nitrogen can be lost from the plant-soil system by one or more of the following processes: leaching; surface runoff; soil erosion; ammonia volatilization; and denitrification. Nitrogen management practices that improve nitrogen efficiency

Nitrogen management aims to maximize the efficiency with which crops use applied N. Improvements in nitrogen use efficiency are associated with decreases in N loss from the soil. Although losses cannot be avoided completely, significant improvements can be realized by applying one or more of the following management practices in the cropping system.

Reduction of greenhouse gas emissions

- Climate Smart Agriculture includes the use of 4R Nutrient Stewardship principles to reduce field emissions of nitrous oxide (N2O) from the application of nitrogen fertilizer. Nitrogen fertilizer is an important driver of nitrous oxide emissions, but it is also the main driver of yield in modern high production systems. Through careful selection of nitrogen fertilizer source, rate, timing and placement practices, the nitrous oxide emissions per unit of crop produced can be substantially reduced, in some cases by up to half. The practices that reduce nitrous oxide emissions also tend to increase nitrogen use efficiency and the economic return on fertilizer dollars.

Reduction of N loss in runoff water and eroded soil

- No-till, conservation tillage and other runoff control measures reduce N loss in surface runoff and eroded soil material.
- The use of daily estimates of soil moisture and crop needs to schedule irrigation reduces the risk of surface runoff and soil erosion.

Reduction of the volatilization of N as ammonia gas

- Incorporation and/or injection of urea and ammonium-containing fertilizers decreases ammonia volatilization because good soil contact buffers pH and slows the generation of ammonia gas from ammonium ions.
- Urease inhibitors temporarily block the function of the urease enzyme, maintaining urea-based fertilizers in the non-volatile urea form, reducing volatilization losses when these fertilizers are surface applied; these losses can be meaningful in high-residue, conservation tillage systems.

Prevention of the build-up of high soil nitrate concentrations

Nitrate is the form of nitrogen that is most susceptible to loss from the soil, through denitrification and leaching. The amount of N lost via these processes can be limited by restricting soil nitrate concentrations, especially at times of high risk. This can be done in many ways, although these are not always cost-effective.

Nitrogen rates: Rates of N application should be high enough to maximize profits in the long term and minimize residual (unused) nitrate in the soil after harvest.

- The use of local research to determine recommended nitrogen application rates should result in appropriate N rates.
- Recommended N application rates often rely on an assessment of yield expectations - these should be realistic, and preferably based on accurate yield records.
- Fertilizer N rates should be corrected for N that is likely to be mineralized from soil organic matter and crop residues (especially legume residues).
- Fertilizer N rates should allow for N applied in manure, in irrigation water, and from atmospheric deposition.
- Where feasible, appropriate soil tests can be used to determine residual soil N.

Soil testing for N

- Preplant soil tests provide information on the soil's N-supply power.
- Late spring or pre-side-dress N tests can determine if and how much additional N is needed.
- New soil test and sampling procedures, such as amino sugar tests, grid mapping, and real-time sensors can refine N requirements.

- Post-harvest soil tests determine if N management the previous season was appropriate.

Crop testing for N

- Plant tissue tests can identify N deficiencies.
- Sensing variations in plant chlorophyll content facilitates variable rate N applications in-season.
- Post-black-layer corn stalk nitrate tests help to determine if N rates were low, optimal, or excessive in the previous crop, so that management changes can be made in following crops.

Precision agriculture

- Variable rate applicators, combined with intensive soil or crop sampling, allow more precise and responsive application rates.

Timing of N applications

- Apply N close to the time when crops can utilize it.
- Make side-dress N applications close to the time of most rapid N uptake.
- Split applications, involving more than one application, allow efficient use of applied N and reduce the risk of N loss to the environment.

N Forms, including slow or controlled release fertilizers and inhibitors

- Slow or controlled release fertilizer delays the availability of nitrogen to the plant until a time that is more appropriate for plant uptake - the risk of N loss through denitrification and leaching is reduced by limiting nitrate concentrations in the soil.
- Nitrification inhibitors maintain applied N in the ammonium form for a longer period of time, thereby reducing leaching and denitrification losses.

N capture

- Particular crop varieties are able to more efficiently extract N from the soil and improve N use efficiency. Breeding of crops for efficient N uptake is in progress.
- Rotation with deep-rooted crops helps capture nitrates deeper in the soil profile.
- Cover crops capture residual nitrogen after crop harvest and recycle it as plant biomass.

- Elimination of restrictions to subsoil root development; subsoil compaction and subsoil acidity prevent root penetration in many subsoils worldwide, promoting build-up of subsoil nitrate concentrations which are susceptible to denitrification and leaching when conditions are suitable.
- Good agronomic practice, including appropriate plant populations and spacing and good weed and pest management, allows crops to produce large root systems to optimise N capture and crop yield.

Water Management

Conservation Tillage

- Conservation tillage optimizes soil moisture conditions that improve water use efficiency; in water-stressed conditions, this improves crop yield per unit N applied.

N fertilizer application method and placement

- In ridged crops, placing N fertilizers in a band in ridges makes N less susceptible to leaching.
- Row fertilizer applicators, such as injectors, which form a compacted soil layer and surface ridge, can reduce N losses by diverting water flow.

Good irrigation management can improve N-use efficiency

- Scheduled irrigation based on soil moisture estimates and daily crop needs will improve both water-use and N-use efficiency.
- Sprinkler irrigation systems apply water more uniformly and in lower amounts than furrow or basin irrigation systems.
- Furrow irrigation efficiency can be improved by adjusting set time, stream size, furrow length, watering every other row, or the use of surge valves.
- Alternate row irrigation and fertilization minimizes water contact with nutrients.
- Application of N fertilizer through irrigation systems (fertigation) facilitates N supply when crop demand is greatest.
- Polyacrylamide (PAM) treatment during furrow irrigation reduces sediment and N losses.

Drainage systems

- Some subirrigation systems recycle nitrate leached from the soil profile and reduce nitrate lost in drainage water.

- Excessive drainage can lead to rapid through-flow of water and N leaching, but restricted or insufficient drainage favours anaerobic conditions and denitrification.

Use of simulation models: Short-term changes in the plant-available N status make accurate seasonal predictions of crop N requirement difficult in most situations. However, models (such as NLEAP and Adapt-N) that use soil, weather, crop, and field management data can be updated with day-to-day changes and thereby improve predictions of the fate of applied N. They allows farmers to make adaptive management decisions that can improve N-use efficiency and minimize N losses and environmental impact while maximizing profitability. Additional measures to minimize environmental impact

Conservation Buffers

- Buffers trap sediment containing ammonia and organic N.
- Nitrate in subsurface flow is reduced through denitrification enhanced by carbon energy sources contained in the soil associated with buffer vegetation.
- Buffer vegetation takes up nitrogen, other nutrients, and reduces loss to water.

Constructed wetlands

- Constructed wetlands located strategically on the landscape to process drainage effluent reduces sediment and nitrate loads to surface water

Observing Climatic Elements

The climatic elements the observation of which is valuable for agricultural purposes can be approached on an idealized threefold scale: (1) microscale observations of small areas for research designed to elucidate basic physical processes; (2) mesoscale climatic networks designed for practicing farmers to improve their operations; and (3) macroscale regional networks intended for weather forecasting and for gathering basic climatic data. Macroscale stations can be further divided into first-order and second-order stations, the number and type of observations different for each. Micrometeorology demands the most elaborate array of measuring devices, while a second-order macroscale station requires the least; in fact, the latter station will measure only five elements: air temperature, rain, snow, humidity, and surface wind. A first-order macrostation will be equipped to measure 16 elements: global radiation, sunshine hours, clouds, net radiation, air temperature, soil temperature, rain, snow, hail, dew, fog, humidity, pan evaporation, pressure, upper air wind, and surface wind. Mesoscale measurements include 10 elements and microscale 27 (three of which are derived from others).

The World Meteorological Organization and the various national weather services are concerned with establishment and improvement of macroscale regional climatic stations, both first-class and second-class. Spaced at least 10 miles (16 kilometres) apart, their value for daily agricultural operations is limited, but they are useful for long-range planning and forecasting. Most parts of North America, Europe, and Australia have adequate networks of these stations, but wide gaps exist in the tropics, polar regions, and arid lands.

The degree day: One weather characteristic of agricultural value is the degree day. This concept holds that the growth of a plant is dependent on the total amount of heat to which it is subjected during its lifetime, accumulated as degree days. Common practice is to use 50° F (10° C) as a base. Thus, if the mean daily temperature for a particular day is 60° F (16° C), then 10 degree days are accumulated for that day on the Fahrenheit scale. The total number of growing degree days required for maturity varies with crop variety as well as plant species. Also, the minimum threshold temperature (the temperature below which the plant is damaged or unable to grow) varies with plants; e.g., 40° F (4° C) for peas, 50° F (10° C) for corn (maize), and 55° F (13° C) for citrus fruits. Where studies have established the number of degree days required for maturity of a given crop, the planting dates can be scheduled for orderly harvest and processing. The system is helpful in selecting crop varieties appropriate to different geographical areas; it also has value in scheduling spray programs and predicting insect emergence.

The growing-degree-day concept has certain weaknesses: (1) it assumes that the relationship between growth and temperature is linear (actually it is not); (2) it makes no allowance for changing threshold temperatures with advancing crop development; (3) too much weight is given to temperatures above 80° F (27° C), which may be detrimental; and (4) no account is taken of the diurnal temperature range, which is often more significant than the mean daily value.

Weather effects: The essence of the weather-agriculture interaction for the farmer lies in wise adaptation of operations to the local climate and in techniques for manipulating or modifying the local environment (microclimate) to minimize weather stresses on plants and animals.

Many of these techniques have been practiced for centuries: seeding and cultivation, irrigation, frost protection, animal shelters, windbreaks, and others are methods of altering the microclimate. The climatic factors and their relation to plant growth in terms of protective techniques are important.

The effects of pollution: Practically all forms of technology exact a certain price in environmental damage; agriculture is no exception. Agriculture in turn is sometimes damaged by undesirable by-products of other technologies.

Air has physical properties and a chemical composition that are vital parameters of life for both plants and animals. Temperature, water vapour, movement, oxygen, and carbon dioxide in the atmosphere have a direct effect on food and fibre production. Air quality is changed by introduction of contaminants into it, and agricultural activities using such air may be affected adversely. Damage to plants by air pollutants is related to meteorological conditions, particularly temperature inversions in the atmosphere.

Air pollution: For more than a century air pollution has affected agriculture. Burning coal and petroleum produce sulfur oxides. Fluorides result from smelting and glass and ceramic manufacture. Rising levels of ammonia, chlorine, ethylene, mercaptans, carbon monoxide, and nitrogen oxides are found in the air. Motor vehicles and growing population produce photochemical air pollution affecting not only the urban concentrations but also the contiguous rural areas. The mixture of pollutants from all sources, including agriculture, has released a host of contaminants into the air, such as aldehydes, hydrocarbons, organic acids, ozone, peroxyacetyl nitrates, pesticides, and radionuclides. The effect of these pollutants on food, fibre, forage, and forest crops is variable, depending on concentration, geography, and weather conditions. Damage to crops by air pollution, of course, brings economic loss as well.

The effects of air pollution on plants and animals may be measured by the following factors: (1) interference with enzyme systems; (2) change in cellular chemical constituents and physical structure; (3) retardation of growth and reduced production because of metabolic changes; (4) acute, immediate tissue degeneration. Pollutants that enter the air from sources other than agriculture and that produce plant response are classified as: (1) acid gases; (2) products of combustion; (3) products of reactions in the air; and (4) miscellaneous effluents.

Acid gases: Acid gases include fluorides, sulfur dioxide, and chlorine. Hydrogen fluoride is extremely toxic to plants; some plants are injured by contact with concentrations of less than one part per billion. The damage apparently occurs initially to the chlorophyll, producing a mottled chlorosis and later killing the cells. Plants vary in degree of tolerance to hydrogen fluoride; usually the plants that accumulate fluoride readily are the most tolerant. Corn is more susceptible than tomato. All plants are most susceptible to fluoride injury during periods of rapid growth.

Sulfur dioxide given off in combustion of oil and coal commonly causes necrosis (cell death) of the leaf. At certain concentrations, sulfur dioxide will affect plants if the stomata (minute pores in the epidermis of a leaf or stem) are open. High light intensity, favourable growth temperatures, high relative humidity, and adequate water supply are conducive to open stomata. Plants that close their stomata at night can tolerate sulfur dioxide much better during that period. Conifers are more susceptible in spring and early summer, when the new needles are elongating. The sulfur dioxide absorbed by the leaf cells unites with water to form a toxic sulfite, but this is slowly oxidized to a relatively harmless sulfate. The toxicity of sulfur dioxide thus is a function of the rate at which it is absorbed by the individual plant; rapid absorption will cause greater injury. Chlorine damage to plants is somewhat rare; its typical symptoms are bleaching and necrosis of the leaf.

Products of combustion: The primary products of combustion are ethylene, acetylene, propylene, and carbon monoxide. Of these, ethylene is known to affect plants adversely; while the others may also do so, it would require higher concentrations of them than typically occur in polluted air. For many years it was observed that illuminating gas (3 percent ethylene) leaking from pipelines caused damage to nearby vegetation. Now, with the use of natural gas, ethylene in the air is derived mostly from certain chemical industries and from automobile exhaust. Greenhouse flowers in metropolitan areas are typically damaged by ethylene. Such injury appears to be caused by excessive speeding up of the life process, thus bringing on damage. Ethylene was first identified as affecting plant life over large areas in the field by its effects on cotton and other plants near a polyethylene factory.

Ethylene, ozone, and peroxyacetyl nitrate are produced as reaction products in the air and are clearly implicated in plant injury. In addition, certain bisulfites and nitrogen dioxide are under suspicion;

there are probably others. Ozone is a major air pollutant affecting agriculture. Damage has been identified in a number of field crops, including spinach, tobacco, fruits, vegetables, forest trees, and ornamentals. Symptoms of ozone toxicity appear as flecks, stipple, streaks, spots, tipburn, and premature yellowing of the foliage; these may be visible only on the upper leaf surface. Peroxyacetyl nitrate and its analogs produce symptoms called silver leaf and leaf banding, which have been observed in the Los Angeles area and elsewhere for many years.

The adverse effects of airborne radioactive contaminants on the agricultural economy at the present time are small.

Air pollution by agriculture: Contributions of agricultural technology to air pollution include pesticides, odours, smoke, dust, allergenic pollens, and trash. The widespread public concern about pesticides makes it imperative that pesticide technology be carefully controlled and that search for better methods be pursued vigorously.

Pesticides: The problem of persistence in pesticides can be highlighted by noting that this attribute exists in a range from moderately persistent (a lifetime of one to 18 months-2,4-D, atrazine); persistent (lifetime up to 20 years-DDT, aldrin, dieldrin, endrin, heptachlor, toxaphene); or permanent (lead, mercury, and arsenic). Presumably, the less persistent types should be more desirable, other things being equal; but those that degrade rapidly, such as the organophosphate insecticides, are extremely toxic and nonselective, which encourages rapid emergence of resistant insects and destroys their natural enemies. Thus, it is apparently not possible to adopt chemicals that function without some drawback or disadvantage.

Whether pesticides are applied by spraying or by surface application, air is the usual medium through which the chemicals move to their intended and unintended targets. Reliable data on how pesticides behave in air, such as distance travelled, are lacking, because adequate monitoring is unavailable. Their chemical and physical nature, method of application, and the atmospheric conditions will influence their concentration and ultimate fate. There is no doubt that pesticides may be transported long distances on dust particles. The rate of removal from air is difficult to predict, but in the long run the chemicals return to earth.

Bibliography

Ashworth S.: *Seed to Seed*, Decorah, Seed Savers Publications, 1991.

Boyer , J.S.: *Measuring the Water Status of Plants and Soils,* Academic Press, N.Y., 1995.

Brady, Nyle C. *The Nature and Properties of Soils,* Eighth Edition. New York: Macmillan, 1974.

Carter, Vernon Gill and Dale, Tom. *Topsoil and Civilization.* Norman: University of Oklahoma Press, 1974..

Chandra, Ramesh and Satish Kumar Singh: *Fundamentals and Management of Soil Quality*, Westville Pub, Delhi, 2009.

Devi, C.R. Sudharmai : *Analytical Procedures in Soil Science and Agricultural Chemistry*, Agrotech, Delhi, 2004.

Dubey, Sarvesh Kumar and Asha Arora: *A Practical Book on Soil, Plant, Water and Fertilizer Analysis*, S.R. Scientific, Delhi, 2010.

Ferentinos L.: *Proceeding of the Sustainable Taro Culture for the Pacific Conference*, Honolulu, HITAHR, 1993.

Golueke, Clarence G., Ph.D. *Composting: A Study of the Process and its Principles.* Emmaus: Rodale Press, 1972.

Gupta, O.P. : *Water in Relation to Soils and Plants : With Special Reference to Agriculture*, Agrobios, Delhi, 2002.

Herminie Broedel Kitchen: *Soils and Crops : Diagnostic Techniques*, Satish Serial Publishing, Allahabad, 2004.

Hopkins, Cyril G. *Soil Fertility and Permanent Agriculture.* Boston: Ginn and Company, 1910.

Jackson, E. : *Crop Management and Soil Conservation*, Biotech Books, Delhi, 2011.

Jana B L : *Water Harvesting and Watershed Management*, Agrotech, Delhi, 2008.

Krasilnikov, N A.: *Soil Microorganisms and Higher Plants.* Y.A. Halperin. Jerusalem: Israel Program for Scientific Translations, 1961.

Kumar, N. : *Breeding of Horticultural Crops : Principles and Practices*, New India Pub, Delhi, 2006.

Majumdar, S.P. and R.A. Singh: *Analysis of Soil Physical Properties*, Agrobios, Delhi, 2008.

Malcolm D. Bolton: *A Guide to Soil Mechanics*, Universities Press, Delhi, 2003.

Nagamani; A ; I K Kunwar and C Manoharachary: *Handbook of Soil Fungi*, I K International, Delhi, 2006.

Naik, M K and G S Devika Rani: *Advances in Soil Borne Plant Diseases*, New India Pub, Delhi, 2008.

Patel, Sheelwant: *Indian Forests : Soil Water and Bio-Environment Conservation*, Pointer, Delhi, 2005.

Rao, A.V. Narasimha: *Fundamentals of Soil Mechanics*, Laxmi Publications, Delhi, 2010.

Rob Jenkins and C.K. Jain: *Advances in Soil-Borne Plant Diseases*, Oxford Book Company, Delhi, 2010.

Schuphan, Werner. *Nutritional Values in Crops and Plants.* London: Faber and Faber, 1965.

Sharieff, Afzal : *Fundamentals of Soil Geography*, Sarup Book Pub, Delhi, 2010.

Sharma, Premjit: *Applied Soil Ecology*, Gene Tech Books, Delhi, 2007.

Singh, S K : *Biotechnology, Plant Propagation and Plant Breeding*, Campus Books, Delhi, 2008.

Turner, Frank Newman. *Fertility, Pastures and Cover Crops Based on Nature's Own Balanced Organic Pasture Feeds.* reprinted from: Faber and Faber, 1955.

Whealy K.: *The Garden Seed Inventory*, Decorah, Seed Saver Publications, 1988.

Index